ENCYCLOPAEDIA OF
OPTICS

ENCYCLOPAEDIA OF OPTICS

Volume 2

Geometrical Optics

Mariyam Noor

ANMOL PUBLICATIONS PVT. LTD.
NEW DELHI - 110 002 (INDIA)

ANMOL PUBLICATIONS PVT. LTD.
4374/4B, Ansari Road, Daryaganj
New Delhi - 110 002
Ph.: 23261597, 23278000
Visit us at: www.anmolpublications.com

Encyclopaedia of Optics

First Published, 2006

ISBN 81-261-2613-2 (Set)

PRINTED IN INDIA

Published by J.L. Kumar for Anmol Publications Pvt. Ltd., New Delhi - 110 002 and Printed at Mehra Offset Press, Delhi.

Contents

Preface

Science is the sermon of today. In every walk of life, since childhood till old age, science plays an important role in the life of man. Uptill some decades back, studied merely as a subject in schools and colleges, Science has assumed the status of an essential part of life today. With the expansion of literacy among the masses and growing awareness, they have woken up from the slumber of ignorance and have started taking up things more sincerely and thoughtfully. It is one of the reasons that our nation is heading towards new goals day after day.

It is a vast subject to study and talk about. For the purpose of ease, Science has been divided into more than two hundred and fifty branches. Specialisation in any branch can be obtained at different institutions and universities, scattered all over the country and abroad.

Apart from other super-specific branches, Physics, Chemistry, Zoology, Botany, Mathematics and Computronics, etc. are more general and closer to daily human life. From the point of view of the cut-throat competition in educational world, a thorough study of the above-mentioned branches, proves very handy, for the new aspirants, especially and for teachers and tutors, generally.

An urgent need was being felt for such a series of books, encompassing the commonly-studied and discussed branches of Science of equal significance and presented in an easy language, understood by all. With God's grace, our dream came true and we are happy, while introducing the dream-series for the lovers of knowledge.

The series consists of many a title, covering various branches of Science, such as Physics, Chemistry, Zoology, Botany, Geology, Agriculture Science, Computer Science, Mathematics and so on. The series is important and a must not only for students but also for teachers. Explained in simple language and aided with graphics and diagrams, everything, between the two ends—title and back cover—of the series, has been presented in a perfect way.

This set of books is devoted to **Optics**, which comprises three volumes: (1) Physical Optics, (2) Geometrical Optics and (3) Optics and Light.

It is hoped that the series would serve both, the students community and the academics, equally and will prove to be a boon for those, aspiring for newer heights. Nevertheless, any suggestions for further betterment of the endeavour will be highly appreciated and solicited.

Editor

1

The Instruments

Camera and Sextant

Photographic Camera : The principle of a photographic camera is that of a convex lens forming a real image on a sensitive plate held at the back of a light-tight chamber formed by canvas bellows with blackened interior. The lens is mounted on a slide so that its distance from the sensitive plate can be altered for bringing to sharp focus the images of objects situated at varying distances.

An adjustable stop is placed in front of the lens so that the amount of light passing through it may be varied. It can be easily shown that the intensity of illumination E of the image is directly proportional to the area of the aperture and inversely proportional to the square of the focal length of the lens, *viz.*'

$$E \propto \frac{\text{area of the aperture}}{(\text{focal length})^2}$$

If d be the diameter of the aperture and f the focal length of the lens, then

$$E \propto \frac{\pi\left(\frac{d}{2}\right)^2}{f^2} = \text{constant} \times \frac{d^2}{f^2}. \qquad \dots (1)$$

The quantity d/f is called the aperture ratio of the lens. It is clear from equation (1) that different lenses with the same value of d/f will give the same brightness of a given object.

The reciprocal f/d of the aperture ratio is called the focal ratio or f number* of the lens. Thus, a lens of 20 cm focal length and 4 cm diameter is said to have a f-number of 5 which means that the diameter of the lens aperture is $f/5$.

Since exposures needed under given conditions may be assumed to be proportional to the square of this number, the apertures are usually indicated in a series such that the exposures required are doubled in passing from one to the next, thus we have the series

$$\frac{f}{2}, \frac{f}{2.8}, \frac{f}{4}, \frac{f}{5.6}, \frac{f}{8}, \frac{f}{11.3}, \frac{f}{16}, \text{ etc.}$$

Speed of a Lens and Depth of Field : The speed of a lens is the measure of the rapidity with which the photographic image is built up. It is obviously proportional to the gain of brightness which as we have seen above is given by

$$E = \text{constant} \times \left(\frac{d}{f}\right)^2 = \frac{\text{constant}}{(f/d)^2} = \frac{\text{constant}}{(f - \text{number})^2}$$

Hence the speed of a lens varies inversely as the square of the f-number. For instance, an $f/2$ lens is faster than $f/4$ lens in the ratio and is more suitable for taking the photographs of faintly illuminated objects and of those which are moving relatively to the camera and hence require a very short exposure.

$$\left(\frac{f}{2}\right)^2 \div \left(\frac{f}{4}\right)^2 = 4:1$$

But it is limit worth noting that with a lens of wide can relative aperture, besides the cumbersome aberrations, points like *O* not accurately in focus give comparatively large circles of confusion like *bb* on the plate P placed. The range of object distances giving reasonably clear images will be small and this is expressed as the lens having a little depth of field.

It is clear that if the lens aperture is reduced then for a given position of *O* the diameter of the circle on the plate will become smaller thus increasing the depth of field, that is, the range of object distances besides improving the definition of accurately focused images. This explains why in a simple box camera with a small aperture and the film placed in the focal plane of the lens, objects very close to the camera lens can produce a reasonably clear picture.

Photographic Objectives : The most important feature of a good camera lens as a rule is that is has a high relative aperture with a large field of view and suffering as it does from the defects of overcome in its construction. The designers of such objectives have therefore, resorted to compromises as regards these corrections that best serve their purpose.

Three types of lenses are in general use :

1. Meniscus lens
2. Rapid rectilinear
3. Triplet anastigmats : Zeiss Tessar.

The above two types of objectives have largely been superseded by recent triplet anastigmats, the most prominent of them being Zeiss Tessar. The chief quality of these lenses is that they give a flat field by actually removing the defects of astigmatism and curvature by making the power of the central flint element equal to the sum of the powers of the crown elements. By placing the concave lens between the two convex ones, the marginal rays can be made to pass through the negative lens so close to the axis that on the whole the triplet is of positive power.

This is a very fast lens and is generally used in the motion picture cameras.

Telephoto Lens : The size of the image of a distant object is proportional to the angle which the object subtends at the camera lens and the focal length of the lens. Hence to obtain a fair-sized image of some distant object, an objective of large focal length is required but this would make the camera itself inconveniently long and unmanageable.

Therefore, photographs of distant scenes are taken with a telephoto lens. It consists of two achromatic lenses, one being convex and the other concave. The latter occupies the normal lens position in the camera box while the former (convex) is placed some distance in front of it. The distance between them is less then the focal length of the convex lens L_1. The chief advantage lies in the fact that while the focal length is long the distance from the back of the objective to its second principal focus is such as can be easily accommodated in a folding camera.

Parallel rays from a very distant object incident on L_1 are brought to focus at F behind the first focus of the divergent lens which is of short focal length. The point F serves as a virtual object for the concave lens which brings them to focus at F_1 where the sensitive plate is placed. If these rays are produced backwards until they cut original beam at P, then it is seen that the combination is equivalent to a single convex lens placed at PN, its focus being at F_1 Without the divergent lens the length of the camera would be approximately FA ; the divergent lens increases it to $F_1 A$ but gives a greatly increased focal length to the combination for the power is decreased to

$$\frac{1}{f_1}+\frac{1}{f_2}-\frac{a}{f_1 f_2},$$

where f_1 and f_2 are the focal lengths of the component lenses and a, the distance between them. Thus the telephotographic camera has an effective length which is much greater than its actual length. Such a system of lenses then enables us to take highly magnified pictures with an ordinary camera.

The focal length of a telephoto lens can be varied at will by altering the distance between its two components. But the advantage thus gained will result to loss of definition for settings other than the standard one for which necessary corrections have been made. Hence it is usually desirable to have telephoto objectives with a fixed focal length.

The Sextant **:** This instrument was devised by Hadley for measuring the angular separation between two distant objects or the altitude of heavenly bodies. It is based on the principle that *the deviation produced in a ray by successive reflections from two inclined mirrors is constant for all angles cf incidence and is twice the angle between the mirrors.*

It consists of a rigid triangular frame being the sector of a circle having an angle of 60°, *i.e.*, one sixth part of the circumference and hence the instrument is called a sextant. There are two plane mirrors A and B, set at right angles to the plane of the frame. The mirror A called the index glass, is the movable mirror and is placed at the centre of the circle. It is fixed to an arm AV which moves over the *graduated* arc PQ and carries a vernier scale V capable of reading upto 10 to 15 seconds. The mirror B, called the horizon glass is, fixed to one of the rigid arms AP. It is half silvered so that if an object is viewed through the telescope T attached to the other arm AQ and directed towards B, two images are in general observed, one seen directly along CBE and the other seen after two reflections along the path *DABE*, when the movable arm is on the zero of the scale, so that the two mirrors are parallel to each other. In this position a distant object is seen simultaneously, the two images exactly coinciding with each other.

To measure the angle subtended at the position of the sextant by two distant points in a vertical line, the sextant is set in a stand with its plane vertical so that the lower of the two points is at the same height as *A*. The telescope is directed to see the lower point through the unsilvered portion of *B* and the movable arm is then rotated until the image of the upper point *H*, seen through the telescope, after two reflections at *A* and *B* coincide with that of the lower point. Suppose *HABE* is the path of rays from the upper point. Let *NAM* and *BM*, the normals to mirrors *A* and *B* in this position meet in *M* then the angle *AMB* (= *a*) between them, is equal to the angle *AFB* between the mirrors. Now by the law of reflection,

$$\angle HAN = \angle NAB = \theta \text{ (say)}$$

and $$\angle ABM = \angle MBE = \phi \text{ (say)}.$$

If β is the angle which the two points subtend at *E*, we have from Δ *AEB*, the exterior

$$\angle HAB = \angle ABE + \angle AEB,$$

or $$2\theta = 2\phi + \beta \qquad ...(i)$$

Similarly, from Δ *ABM*,

$$\theta = \phi + \alpha,$$

where α is the angle between the mirrors *A* and *B* and hence also between their normals.

From (*i*) and (*ii*) $\beta = 2\alpha$, which means that the angle which the two points subtend at the observer's eye at *E* is double the angle between the mirrors *A* and *B*.

When $\alpha = 0$, $\beta = 0$, that is, *A* and *B* are parallel and the direction of the arm *A V* is then along *AQ* parallel to the fixed direction *BF* of the horizon glass. If the zero of the main scale coincides with the zero of the vernier scale in this position, α is read directly on the arc. Actually, however, the graduations on the arc *PQ* are doubled. That is instead of 60 divisions of 1° each there are 120 divisions so that instead of a, it is the required angle $\beta = 2\alpha$, which is read directly on the scale.

Determination of the Height of a Distant Object : The angle which the distant object *AB,* say a pole, subtends at a point *C* at the same level as the foot *B* is determined with the help of the sextant. If h is the required height and z (= $x + y$) the distance of the pole from C and θ_1 the angle subtended by h at *C,* then $h = z \tan \theta_1$.

If the object is inaccessible, the angles θ_1 and θ_2 which it subtends at two different positions *C* and *D* distance x apart, are determined. Then we have

$$X + y = h \cot \theta_1 \qquad \ldots(iii)$$

and

$$y = h \cot \theta_2 \qquad \ldots(iv)$$

On subtracting (*i*) from (*iii*) $x = h$ (cot θ_1 – cot θ_1),

whence

$$h = \frac{x}{\cot \theta_1 - \cot \theta_2} \qquad \ldots(v)$$

Knowing x, θ_1 and θ_2, h can be calculated. It is assumed that the base line is perpendicular to the object.

Measurement of the Altitude of the Sun : To find the altitude of the sun at sea, the horizon being definite is taken as one object, the sun itself being the other. On land the horizon line is not definite so an *artificial horizon is* formed by a pool of mercury contained in a flat vessel. Since it is highly reflective and being heavy it is not liable to be easily disturbed by the wind. The angle subtended at the eye of the observer by the sun and its image in the mercury is measured. It is obviously equal to twice the altitude of the sun.

Microscopes, Telescopes and Eye-pieces : *The Visual Angle and Visual Acuity* : The apparent size of any object as seen with the unaided eye depends on the angle it subtends at the eye. It is called the visual angle. The retinal image becomes larger and larger as the object is brought closer to the eye on account of the increase in this visual angle.

Three positions of the same object are shown and we find that the size of the retinal image increases as the object approaches the eye,

$$O'C > O'B' > O'A'.$$

But there is a limit as to how close an object can be brought to the eye if the latter is still to produce a sharp image without over strain. For a normal eye this distance of most distinct vision is taken as 25 cm, and the limiting angle which two point objects at this distance must subtend at the eye in order that they may be distinguished as separate is between one to two minutes and is called the visual acuity or the resolving power of the eye.

The purpose of the magnifying instruments like microscopes and telescopes is to increase the range of the eye by forming a virtual image of the object which subtends a much greater visual angle. The eye sees the image instead of the object directly.

The magnifying power of an optical instrument measures the ratio of the visual angle of the image as seen through the instrument to that of the object as seen by the naked eye.

Magnifying Glass : A simple convex lens of short focus is sometimes used as a magnifier. The object AB, to be examined is placed close , to the eye and the short focus convex lens is interposed so that the object is within the principal focus of the lens and at a distance u from it. Its virtual image is formed at the distance of distinct vision (D). Assuming the eye very close to the lens, the image will also be at a distance D from the lens. Let α be the angle which the object AB subtends when at this distance and β the angle subtended by the image $A'B'$ at the eye.

Then for the lens E, $\frac{1}{v} - \frac{1}{u} = \frac{1}{f}$, where f is its focal length.

Since *conventionally*, v and u are both negative, and $v = D$,

$$-\frac{1}{d} + \frac{1}{u} = \frac{1}{f},$$

$$\frac{D}{u} = 1 + \frac{D}{f} \qquad ..(a)$$

By definition, the magnifying power is given by

$$M = \frac{\text{angle subtended by image}}{\text{angle subtended by the object placed at the distance of distinct vision}}$$

$$= \beta / \alpha.$$

Now, $$\beta = \frac{h'}{D} \text{ and } \alpha = \frac{h}{D},$$

$$\therefore \qquad M = \frac{\beta}{\alpha} = \frac{h'}{h} = \frac{D}{u}, \quad ...$$

[by similar Δs $A'EB'$ and AEB]

$v = D$

whence by equation (a) above $M = 1 + \frac{D}{f}$ ...(2)

$v = D$

With the image formed at the distance of distinct vision, the eye has to exert its maximum accommodative power. For perfectly relaxed normal eye the far point is at infinity so that the object must be placed at the focus of the lens. In this case,

$$\beta = \frac{h}{f} \text{ and as before }, \alpha = \frac{h}{D}$$

$$\therefore \quad \underset{v \to \infty}{M} = \frac{\beta}{\alpha} = \frac{D}{f}. \qquad \text{... (3)}$$

The magnifying power is, therefore, larger if the image is formed at the distance of distinct vision. But magnifying glasses have generally short focal lengths so there is not much difference between the two expressions and the simpler expression $\frac{D}{f} = \frac{25}{f}$ is commonly used in labelling the power of the magnifiers.

Astronomical Telescope : A refracting type astronomical telescope consists of a convergent object glass of long focal length and a convex eye-piece of short focal length, mounted on the same axis at the opposite ends of a metallic tube. The eye-piece can be slided in or out as desired.

Rays from a distant object are brought to focus in the focal plane of the object glass forming a real image. The eye-piece is used to magnify this real image and to form a virtual image at any distance between 25 cm and infinity.

We have already noted that for a normal-sighted eye objects are viewed most easily by parallel rays, so we shall assume that rays become parallel after emerging from the eye-piece. The telescope is then said to be in normal adjustment.

The optical path of two bundles of rays through the telescope, in this case the rays *AAA* coming from one part and *BBB* from the other part form the inverted real image *ba* in the second focal plane of the object glass. If this is also the first focal plane of the eye lens, these rays spread out from *ab* and eventually on emergence from the eye lens come out as two bundles of parallel rays. When these emergent bundles fall on the eye, the final inverted and magnified image is seen at $A'B'$.

The magnifying power of the telescope is defined as the *ratio of the angle subtended at the eye by the final image A'B' to the angle subtened at the eye by the object itself.* Let the object subtend an angle α at the objective then it would subtend approximately the same angle at the eye, as the distance between the eye and the objective is negligible compared to the distance of the object from the telescope. Let the angle subtended at the eye by the final image be β.

By definition, magnifying power. (M) = β / α.

Now $\beta = \angle \beta Ea = \frac{ab}{f}$, where f is the focal length of the eyepiece and $\alpha = \angle AOB = \angle bOa = \frac{ab}{F}$, where F is the focal length of the objective.

$$\therefore \qquad M = \frac{\beta}{\alpha} = \frac{ab}{f} \div \frac{ab}{F} = \frac{F}{f}. \qquad \text{... (4)}$$

Hence, the *magnifying power of a telescope in normal adjustment is the ratio between the focal length of the objective and that of the eyepiece.*

When a telescope is employed to see near objects as in our laboratory experiments, then it is not in normal adjustment and the final image may be caused to be at any desired position convenient to the observer.

E represent the objective and the eye lens of a telescope respectively, *AB* is the object at a distance *u* from the objective, *ba* its image in the objective and $B_1 A_1$ the final image as seen through the system at a distance *v*, say, from the eye lens. The magnifying power is given by

$$m = \frac{\beta}{\alpha} = \frac{\text{angle subtended by the image at the eye}}{\text{angle subtended by the object seen directly by the eye}}$$

If the eye is placed close to the eye lens then assuming that the length of the telescope tube is small as compared to u, the angle subtended by the image at the eye $(\beta) = \frac{h}{u_1}$ by the object $(\alpha) = \frac{h}{x}$.

$$\therefore \quad M = \frac{\beta}{\alpha} = \frac{h}{u_1} \div \frac{h}{x} = \frac{x}{u_1},$$

where x is obtained by the relation (using numerical values),

$$\frac{1}{x} + \frac{1}{u} = \frac{1}{f} \text{ for the objective,}$$

whence $\quad x = \frac{uF}{u-F}.$

u_1 is obtained by the relation, $-\frac{1}{v} + \frac{1}{u_1} = \frac{1}{f}$ for the eye lens,

whence $\quad u_1 = \frac{vf}{u+f}.$

Substituting these values of x and u_1 in the above equation, we have

$$M = \frac{uF}{u-F} \times \frac{v+f}{vf}. \quad \text{...(5)}$$

Special Cases

1. When the object is at a distance u and the image is formed at the least distance of distinct vision ($v = D$), equation (5) may be written as

$$M = \frac{uF}{u-F} \times \frac{D+f}{Df}.$$

2. When the object is at a distance u and the image is at infinity ($v \to \infty$), equation (5) may be written as

$$M = \frac{uF}{u-F} \cdot \frac{1+\frac{f}{v}}{f}.$$

or
$$\underset{v\to\infty}{M} = \frac{u\,F}{u-F} \times \frac{1}{f} = \frac{u}{u-F} \cdot \frac{F}{f}.$$

Since $\frac{u}{u-F}$ is greater than 1, the magnifying power for objects at finite distances is greater than F/f, the magnifying power in *normal* adjustment.

3. When the object is at infinity ($u \to \infty$) and the image is at the least distance of distinct vision $v = D$, equation (5) may be written as

$$M = \frac{F}{1-\frac{F}{u}} \times \frac{D+f}{D}$$

or
$$\underset{u\to\infty}{M} = \frac{F}{f} \cdot \frac{D+f}{D}.$$

4. When both u and v are infinite, the telescope is in normal adjustment, then equation (5) gives

$$M = \frac{F}{1-\frac{F}{u}} \times \frac{1+\frac{f}{v}}{f}$$

$$= \frac{F}{f}, \text{ a value obtained already.}$$

The Exit Pupil and the Entrance Pupil : We find that all rays which pass through both the lenses, pass through the area, which is in fact the image of the apertura of the object glass in the eye lens. This area is called the exit pupil or the eye ring or the Ramsden's circle and the point where the area intersects the axis of the telescope is called the eye point.

In telescopes object glass is the aperture stop and is, therefore, the entrance pupil. The eye ring is in fact the image of the object glass and all rays which enter the object glass must emerge through the eye ring. The exit pupil is, therefore, the place where the eye must be situated to see the largest range of objects. If eye is placed on either side of the eye point, the outer bundle of rays would not fall on the eye at all and so the outer parts of the object would not be visible. The instrument is generally designed with its eye ring smaller than the eye pupil so that all the light from the object glass enters the eye. It is important, therefore, to know the diameter KJ of the eye ring. Let it be d and that of the object glass be D.

Now, $KJ = LN\ \theta \times f$, where $\theta = \angle LbN$ and f is the focal length of the eye-piece. Also $\theta = D/F$, where F is the focal length of the objective.

$$\therefore \qquad KJ = d = \frac{D}{F} \times f = \frac{D}{F/f} = \frac{D}{M}. \qquad \ldots (9)$$

That is, the diameter of the eye ring varies *directly* as the diameter of the objective and *inversely* as the magnifying power.

We shall see later in the chapter on the resolving power of optical instruments that for good definition of the image and large resolving power of the instrument the use of objectives with large apertures is a great advantage. Consequently, increase in the diameter of the eye ring may be limited by increasing the magnifying power of the telescope. But again the magnifying power should not exceed that at which the size of the exit pupil matches that of the eye. This particular value is called the normal magnifying power.

It is at this value that the illumination is maximum for the field of view is seen by full pencils and it is the largest.

The diameter of pupil of the eye is generally taken to be 2.54 mm so that the size of the exit pupil should be less than 2-54 mm but it should not be less than 085 mm, for below it the image will not be bright enough.

Then, $\frac{Df}{F} \ngtr 2.54$ mm; if $D = 12.70$ cm, $\frac{f}{F} \ngtr \frac{1}{50}$ or $\frac{F}{f} \nless 50$.

If $\frac{Df}{F}$ is still smaller, $\frac{F}{f}$ can be made still larger.

In Yerkes' refracting telescope the diameter D of the aperture of the objective is 101.6 cm (40 inches)

so that if $\frac{Df}{F} \ngtr 2.54$ mm, $\frac{F}{f} \nless 400$.

Also if $\frac{Df}{F} \nless 0.85$ mm, then $\frac{F}{f} \ngtr 1200$,

so that the minimum and maximum magnifying powers are 400 and 1200 respectively.

From this discussion, it is obvious that if F, f and D are subject Df mainly to the restriction that $\frac{Df}{f} \ngtr 2.54$ mm, that is, there is no upper limit to the magnifying power of the telescope. We find, however, that beyond a certain point nothing is gained by increasing magnification, the image becomes larger but reveals no fresh details. The focal length of the objective and its aperture both should be increased.

We shall come to this subject again in the study of the resolving power of the optical instruments.

Example : *A spectrometer telescope objective of diameter 3.5 cm and focal length 35 cm is fitted with an eye-piece of focal length 4 cm. Find the magnifying power of the telescope and the diameter of Ramsden's circle.*

Here F = 35 cm, f = 4 cm, D = 3.5 cm.

The magnifying power, $M = \frac{F}{f} = \frac{35}{4} = 8.75.$

The diameter of Ramsden's circle, $d = \frac{D}{F/f} = \frac{D}{M}$

$$= \frac{3.5}{8.75} = 0.4 \text{ cm.}$$

Reflecting Telescope : Most of the large astronomical telescopes of the world today employ concave paraboloidal mirrors in place of achromatic glass objectives. They have several advantages over the refracting type.

1. A concave reflector is free from chromatic aberration and also spherical aberration if paraboloidal in form, thus permitting a very high relative aperture and so giving a very high resolving power. This is a point of great importance in the study of stars.
2. Large aperture lenses of pure optical glass of sufficient homogeneity are difficult to be manufactured. Furthermore, an achromatic objective must be constructed of at least two lenses which would require the grinding and polishing of four surfaces while for a mirror only a single surface requires finishing. The objective of the world's largest refracting telescope at Yerkes is only 101.6 cm (= 40") in diameter.
3. Greater stability of the telescope is obtained by having the largest and the heaviest optical part at the bottom.

The movements are electrically controlled and this has made the observations easier and more accurate.

The object glass is a concave mirror C which reflects the parallel rays coming from the distant object to a small plane mirror M at an inclination of 45° to the axis so that the real image is formed in its front at $a'b'$. It is viewed through the eye-piece E whose axis is at right angles to the axis of the instrument and the final image is formed at $A'B'$. The magnifying power is the ratio of the angles $a'Eb'$ and bCa, taking the latter as the angle which the object would subtend at the eye if viewed directly.

$$\therefore \qquad M = \frac{\angle a'Eb'}{\angle aCb} = \frac{a'b'}{f} \div \frac{ab}{F} = \frac{F}{f}, \qquad \ldots [\because ab = a'b']$$

where F and f are the focal lengths of the concave mirror and the eyepiece respectively.

Other reflecting telescopes based on the same principle have been designed by Herschel, Cassegrain and Gregory. In Herschel's telescope the concave reflector is placed obliquely to the tube and the image is formed in its focal plane near the edge of the tube and viewed directly through an eye-piece. In Cassegrain's type, the light reflected from the concave mirror converges and meets a small convex mirror m before reaching the focus at F. The focal length of the convex mirror is such that a real image is formed just outside a hole at the vertex of the concave mirror, where it is viewed by an eye-piece. Such an arrangement considerably increases the effective focal length of the objective and so gives a greater magnifying power to the telescope. The aperture of the convex mirror is of such dimensions that it just receives all the rays reflected from the concave mirror to F. The diameter of the hole at the centre of the concave mirror should not be greater than the aperture of the convex mirror for otherwise some of the incident light would enter directly into the eye-piece.

Gregory's type is similar to Cassegrain's but the small

mirror is concave instead of convex to the light so that its vertex must be outside the focal length of the large concave mirror.

The Mount Palomer Telescope : The optical features of the great 5.08 m. aperture reflecting telescope. It is of the Cassegrain's type. With the large concave mirror C parallel rays entering the telescope would be brought to focus at *F*. But before reaching *F* the converging bundle of rays is reflected by a convex mirror *m* down the axis and passing through a hole in the objective comes to focus at *S*, a point just outside the hole so that *F* and *S* are conjugate foci for the convex mirror *m*. For visual observations the image *S* falls in the first focal plane of the eye-piece, so that the final image is at infinity. This image is inverted since the second image at *S* is erect with respect to the first image at *F* and this first image is inverted with respect to the object.

If the image is to be photographed, at the point *S* is located a photographic plate or the slit of a spectrograph. The mirror has the light-gathering power of a million eyes and it may be regarded as a giant camera which has enabled astronomers to discover objects a thousand million light years away from the earth. This method has the advantage that the plates can be exposed for hours, if necessary, to the object to be studied and thus enabling permanent records to be made.

It excels in performance the 2.54 m Mount Wilson reflecting telescope and has opened up a view beyond our own stellar system towards hazy patches very much like our own milky way. They are known as extra Galactic nebulae—independent stellar systems some where in the infinite Universe. Within the orbit of practical research thus came all aspects of stellar energy, building up and breaking down of elements in space, the problem of abundance of various chemical elements in stars and the investigation of the theories of the Universe.

The telescope has a focal length of 72.24 m and is located in the dome of California observatory which is 41.76 m in diameter and 41.15 m high. The revolving part alone weighs 1000 tons but is easily controlled and moved by comparatively diminutive electric motors.

Compound Microscope

The subject of microscopy is gaining more and more importance in recent times and greater magnifications and resolving powers have become necessary. In modern work, microscopes, with electronic beam as a source of light, are being employed, thus improving the technique of microscopy considerably. We shall now discuss the principle of a simple form of compound microscope. To obtain very large magnifications and get rid of the several system a compound microscope is generally employed for the study of minute substances.

In its simplest form a compound micrpscope, consists of two convergent lenses, one of very short focus called the objective *O* and the other of somewhat longer focus called the eye-piece *E*. The object *AB is* situated just outside the focus of the objective which forms its real image at *ab*. This image becomes the object for the eye-piece which acting as a magnifying glass forms a large virtual image at *A'B'* anywhere between the near and the far points of the eye. The overall magnifying power *M* of the instrument will clearly be given by the product of linear magnification *Me* due to the objective and the magnifying power *M*, of the eyepiece,

i.e., $$M = M_0 \times M_e.$$

The linear magnification due to the objective is given by

$$M = \frac{v}{u},$$

where v is the distance of *ab* from *O*,

$$\therefore \qquad M = \frac{v}{u}\left(1 + \frac{D}{f_e}\right) \qquad \ldots \left[\because M_e = 1 + \frac{D}{f_0}\right]$$

Also $\frac{v}{u} = 1 - \frac{v}{f_o}$, for the objective, so that when f_0 is small, this may approximately be taken as $-v/f_o$. The eye-piece of focal length *fe* acting as a magnifying glass, gives the magnification M_1 approximately equal to D/f_e. Hence dropping negative sign, we have

$$M = M_0 \times M_e = \frac{v}{f_0} \times \frac{D}{f_e}$$

The system of lenses generally comprising the objective of a microscope is a very complex one and replacing it by principal planes as the value of M_e may be written as $\frac{L}{f_0}$ where L is the distance bF_2 of the image ab from F_2, the second focal point of the objective and $f_o = H_2F_2$, its focal length.

$$\text{Hence} \qquad M = \frac{L}{f_0} \cdot \frac{D}{f_e} \qquad \ldots(10)$$

L is called the optical length of the microscope and is generally taken as 18 cm. Since D for normal eye =25 cm,

$$M = \frac{18 \times 25}{f_0 \times f_e}$$

$$= \frac{450}{f_0 \times f_e} \qquad \ldots(11)$$

It should be noted that this expression for magnifying power is independent of the nearness of the observer to the eye lens.

In practice the objectives are specified by their focal lengths, 1. 27 cm ($= \frac{1}{2}''$) objective and 0.635cm($= \frac{1}{4}''$) objective, etc., and the eye-pieces by their magnifying power so that the magnifying power of the microscope can be readily calculated.

The optical arrangement of a compound microscope generally employed for examining small particles or slides. It is provided with a rotating nose-piece to which are attached three objectives of different magnifications, one of them being the oil immersion objective. By rotating the nose-piece, any one of the objectives may be set in proper alignment with the eyepiece.

The object or the slide to be examined is placed on the stage S of the microscope with a hole in the centre to allow light to pass through it. Immediately below the stage is a thick convex lens C called the condenser. Below this is a mirror M, plane on one side and concave on the other, capable of rotating about a horizontal axis. Light rays from a strong source of light are reflected by the mirror so as to fall on the condenser, which in turn focuses them on the object to be examined.

The Exit Popil and Magnifying Power : As in the case of a telescope, the exit pupil of a microscope is the image of the aperture of the objective formed by the eye lens and it is the place where the eye must be placed to see the greatest *range* of the object. From the course of rays it is clear that KJ represents the exit pupil of the microscope. If r be its radius, then

$$2r = KJ = MN = 2f.\ \tan \alpha_{1,} \qquad \ldots(a)$$

where f_e is the focal length of the eye lens and

$$a_1 = \angle PDO = \angle NDE.$$

Since $\angle PCO(=\alpha)$ is the object slope angle for the objective and α_1 is the corresponding image slope angle, then by the sine condition we have

$$\mu h \sin \alpha = \mu_1 h_1 \sin \alpha_1,$$

where h_1 is the linear length of the image formed by the objective of a small axial object of corresponding length h.

It follows that M_o, the magnification due to the objective is given by

$$M_0 = \frac{h_1}{h} = \frac{\mu \sin \alpha}{\mu_1 \sin \alpha_1} \quad ...(b)$$

The quantity $\mu \sin \alpha$, that is, the product of the refractive index of the object medium and sine of the slope angle of the outermost ray from an axial point on the object is called the numerical aperture (*N.A.*) of the objective of a microscope. The name was given by Abbe (1840—1905). It is important chiefly in connection with the brightness of image, depth of focus and the resolving power of the microscope.

Now equation (*b*) may be written as

$$M_o = \frac{N.A.}{\sin \alpha_1}, \text{ since } \mu_1 = 1.$$

As α_1 is small we may put $\sin \alpha_1 = \tan \alpha_1$.

$$\therefore \quad M_o = \frac{N.A.}{\tan \alpha_1}.$$

Since the magnifying power of the microscope.

$$M = M_o \times M_o = M_o \times \frac{25}{f_e},$$

$$\therefore \quad M = \frac{N.A.}{\tan \alpha_1} \times \frac{25}{f_e} = \frac{25\,N.A.}{r} \quad ...$$

[from (*a*) above] ...(12)

$$\therefore\ r, \text{ the radius of the exit pupil} = \frac{25\,N.A.}{M} \text{ cm} \quad ...\ (13)$$

Now, for an eye examining small objects in average illumination the diameter of the pupil is 2.5 mm and its radius=1.25 mm. If the optical system is such that the pupil of the eye is completely filled With light emerging from the eye-piece, the magnifying power is said to be normal.

$$\therefore \qquad \text{normal } M = \frac{25\,N.A.}{0.125} = 200\ N.A. \qquad \ldots(14)$$

In actual working, however, a magnifying power much greater than the normal is needed even at the cost of reducing the diameter of the eye-ring as will be explained in the discussion of the resolving power of a microscope. If magnifying power is less than the normal, the radius of the exit pupil exceeds that of the eye pupil and the whole beam of light is not utilized. In this case the actual brightness of the images remains the same as for normal magnifying power but the numerical aperture which is actually used is reduced and there is, as we shall see later, a consequent loss in the resolving power.

Now let us find the magnifying power of a compound microscope when the observer's eye is at the exit pupil and the image is at the least distance of distinct vision.

Let E_1 be the aperture of the objective and E_2 the image of this aperture produced by the eye-piece, *viz.*, the exit pupil. All the rays from the object which enter the microscope must pass through E_2. Hence, if an eye is placed close to the exit pupil with its pupil larger in size than E_2, the image is seen under the best conditions.

Let p, q and x be the distances as indicated and v be the distance of the final image from the eye lens then when $v = D - q$, the magnifying power is given by

$$M = \frac{h_2}{D} \div \frac{h_2}{D} = \frac{h_2}{h_1} \times \frac{h}{h_1}$$

$$= \frac{D-q}{p-x} \times \frac{f_o}{u+f_o}.$$

Since $\frac{1}{q}+\frac{1}{p}=\frac{1}{f_e}$ and also $-\frac{1}{D-q}+\frac{1}{p-x}=\frac{1}{f_e}$, we have

$$\frac{p+q}{pq}=\frac{1}{f_e} \text{ and } \frac{D-q}{p-x}=1+\frac{(D-q)(p+q)}{pq}.$$

$$M=\left[1+\frac{(D-q)(p+q)}{pq}\right]\left(\frac{f_o}{u+f_o}\right). \qquad ...(15)$$

Ultra-microscope

This is an optical arrangement for viewing extremely minute particles of matter, say, some colloidal suspension although no image of the particles is formed and their size and shape etc. cannot be ascertained. The suspension to be examined is contained in a vessel *V* with an optically worked glass window on one side and at the top. In this case the illumination is not direct as in the compound microscope, but a very thin beam of intense light from an arc lamp *S* is allowed to enter the solution horizontally through the side glass window in *V*. It is brought to Principle of ultra-focus on the axis of the microscope, scope *M* by a convex lens *L*. The suspended particles falling in the path of the incident beam scatter light and when focused from above by the microscope held vertically they can be observed only by scattered light against a comparatively dark background, called the dark field illumination.

In this way even those suspended particles can be seen which are far too small to be made visible by ordinary means of illumination. Indeed we can see down to those of which the diameter is 10^5 mm or 1/10 of the wavelength of sodium light. In order that particles should scatter enough light in a direction normal to the incident light, besides intense illumination there should be sufficient difference in the optical refractivities of the liquid and the suspension. To study details ultra-violet light is employed.

By this arrangement Brownian movements can be easily examined. A small metallic chamber containing smoke has a high power convex lens on one side and an optically worked glass window at the top. Parallel light from an arc lamp is focused by the side lens to a point in the chamber on the axis of the microscope very close to its plane glass top through which the scattered light is viewed by means of the microscope. Very queer quivering movements of the smoke particles shining like starry clusters in the dark sky are easily visible.

These movements are due to the random kicks received by the visible smoke particles from the invisible air particles and are a striking confirmation of the kinetic theory of gases. They are known as Brownian movements because they were originally observed by the British botanist Dr. Robert Brown in the study of the plants.

There are alternative arrangements for obtaining dark field illumination. In these types the light is incident upon the object at angles such that it does Not pass by transmission or ordinary specular reflection directly into the objective, that is, the only light to enter must be that which has been diffractor scattered by the object itself. For low power objectives this is achieved by inserting below the sub stage condenser a circular opaque disc with its centre on the axis of the microscope ; its diameter is such that it completely obstructs the central rays which would otherwise enter the objective.

For high power oil immersion objective the improved type of dark field illumination is of the reflecting type which makes use of a cardioid lens. The illuminating rays are reflected first at the spherical reflector *S* and then at the surface of the cardioid *C* and fall on the object at much greater obliquity than in the previous case. By this device both chromatic and spherical aberrations are removed and the condenser becomes truly aplanatic.

Defects of Objectives and Eye-pieces and their Removal : In discussing the simple theory of a telescope and a microscope single lenses were used as objectives and eye-pieces but we have noticed already that such lenses suffer from various defects of spherical and chromatic aberrations and a limited field of view. Hence in actual practice instead of using a single lens for each a suitable combination of lenses is employed.

Objectives :

(i) *Telescope Objective :* We have described how the chromatic aberration of a telescopic objective can be minimised by using combination of a convergent lens of crown glass and a divergent lens of flint glass cemented together by canada balsam. The combination is as a whole plano-convex with convex side facing the parallel rays from the distant object so as to remove spherical aberration.

(ii) *Microscope Objectives :* High quality microscopes are equipped with short focus objectives consisting of co-axial lens systems which make them free from chromatic and spherical aberrations even when a large aperture is used to admit a great deal of light. The necessary corrections for spherical aberration etc. are made by making use of aplana tic surfaces and those due to chromatism by suitable achromats of crown and flint shows Abbe's homogeneous immersion objective. The first lens L_1 of the objective is a pianohemispherical convex lens and the space between this and the object to be examined is filled with cedar wood oil of the same refractive index as that of glass. The advantage of this method of *immersion* is that the divergence of rays emerging from glass side into air is eliminated so that the objective receives more rays from the object than it would, if oil were not used. For in the absence of the oil some of the rays which strike the objective when the oil is present, emerging from the side into the air would be bent outwards so far as to miss the objective. A wide angled cone of rays proceeding from the object. Abbe's homogeneous *A* which is one of the aplanatic points immersion objective of surface *S* of L_1, after refraction from *S* will appear to proceed from A_1. A second lens L_2 is placed above it so that A_1 is the centre of curvature of its lower surface S_1.

Thus the rays are refracted only at the second surface S_2 of L_2 It is also arranged that A_1 and A_2 are aplanatic points with respect to this surface. We have thus a wide-angled beam originating from *A* being converted into a beam of smaller angle coming from A_2 and making its entry into the microscope. In order to correct for chromatic aberration two compound lenses L_3, and L_4 of crown and flint glasses are placed above L_2. Such a system of lenses satisfies Abbe's sine condition so spherical aberration and coma are both removed even when a wide-angled pencil of rays is used. The achromatic doublets produce a colourless image in front of the eye-piece.

Various Eye-piece

The Eye-piece : If a single convex lens is used as an eyepiece, besides the effects of chromatic and spherical aberrations it suffers from a defect arising from an entirely different cause. When the eye is placed close to the eye lens, the rays from the peripheral parts of the image *I* of a distant object formed by the objective *O* would cross over the eye lens and thus fail to enter the eye along with the paraxial rays. Thus the plane of the field near the axis will alone be seen and the field of view would become limited.

Hence to enable the marginal rays to pass through the eye-lens or in other words to increase the field of view another lens F is introduced between the eye lens and the objective, say, in the plane of the image I. This deflects the outer rays towards the principal axis so that after passing through the eye lens these rays enter the pupil of the eye simultaneously with the central rays thus enlarging the field of view. F is called the *field lens, F improves the field* of view and E the *eye lens*, F improves the field of view. And E acts as a magnifier. The combination of E and F is termed the **compound eye-piece** or **ocular.** The two lenses are so designed that the combination is free from spherical and chromatic aberrations. There are several types of oculars used with various optical instruments. They consist of lenses of appropriate focal lengths and radii of curvature and are placed at a suitable distance from each other.

Kellner's Eye-piece : It consists of two plano-convex lenses of equal focal length and separated by a distance equal to the focal length of either. They are placed with their convex surfaces turned towards the incident light. The condition for achromatism.

$$a = \frac{f_1 + f_2}{2} \text{ is fully satisfied.}$$

Since $f_1 = f_2 = f$, say, so that $a = \dfrac{f + f}{2} = f$.

the focal length of either. This condition simply means that red and blue images although of different sizes and formed at different positions subtend the same angle at the eve causing the sensation of whiteness due to overlapping. Spherical aberration is also reduced as the deviation is now spread over four refracting surfaces. In order that the rays must emerge from the eye lens E as a parallel beam for the final image to be seen at infinity before striking E they must have come from its first focal plane. But as the distance between the field lens and eye lens is f_r the focal length of either, the principal plane of the field lens coincides with the first focal plane of the eye lens. Hence the objective must form a real image in the principal plane of the field lens. This will make no difference in the dimensions of the image finally seen since no further magnification is produced by the field lens.

Merits and Demerits

(1) This eye-piece has a very wide field of view and is suitable for use with microscopes.

(2) Since the magnification is entirely produced by the eye lens the marginal parts are disproportionately magnified thus producing distortion.

(3) There is another undesirable feature that the image I and the surface of the field lens are simultaneously in focus, so that any dust particles or a smear on the surface of F or any flaw in its material would be magnified by the eye glass and would seriously interfere with the visibility of the image.

Ramsden's Eye-piece : This is generally used in telescopes and spectrometers and is designed to enable us to measure the dimensions of the image by the aid of a scale or cross-wires in the eye-piece. It consists of a combination of two plano-convex lenses of equal focal length separated by a distance two-thirds of their common focal length and placed beyond the image formed by the objective. The lenses are placed with their plane faces outwards so that spherical aberration effects are reduced. Since the distance apart of the lenses is only $\frac{2}{3}$ of their common focal length instead of f on account of the disadvantage referred to in Kellner's eye-piece, the condition for achromatism is not quite satisfied but the departure from perfect achromatism is not large. Chromatism is sometimes further reduced by using an achromatic lens for each component.

The spherical aberration is reduced by sharing the deviations as equally as possible among the four refracting surfaces. This also eliminates coma to some extent. As narrow pencils pass through the eye-piece, this also helps in reducing the radius of the circle of least confusion due to spherical aberration and astigmatism.

The course of rays through a Ramsden's eyepiece. To examine the image without strain the rays leaving E must be parallel and consequently before striking E must have proceeded from A_1 B_1 in its focal plane. A_1B_1 is the image formed by the field lens of the real image AB formed by the object glass at its second focus.

$\therefore\ EF = \frac{2}{3} f$, the distance from F to $A_1B_1 = \frac{f}{3}$ numerically,

i.e., $v = -\frac{f}{3}$ for the field lens.

Since $\frac{1}{v} - \frac{1}{u} = \frac{1}{f}$,

$$\therefore \quad -\frac{3}{f} - \frac{1}{u} = \frac{1}{f},$$

whence $\frac{1}{u} = -\frac{3}{f} - \frac{1}{f} = -\frac{4}{f}$, or $u = -\frac{f}{4}$.

That is, the real image AB is in front of the field lens. This is the position at which cross-wires must be placed and at which real image due to objective is formed. Hence the eye-piece is sometimes known as positive eye-piece. It magnifies equally both the image at AB and the cross-wires or the transparent micrometer scale placed there. Therefore, it can be easily focused and enables us to take accurate measurements of the size of the image and there is very little distortion in it.

The equivalent focal length F' of the compound eye-piece is given by

$$\frac{1}{F} = \frac{1}{f_1} + \frac{1}{f_2} - \frac{a}{f_1 f_2}$$

$$= \frac{1}{f} + \frac{1}{f} - \frac{2f/3}{f^2} = \frac{2}{f} - \frac{2}{3f} = \frac{4}{3f}.$$

$\therefore$ $F' = 3f/4.$

The equivalent lens of this focal length should be placed $f/2$ *behind* the field lens.

The magnifying power of a telescope fitted with Ramsden's eye-piece is increased somewhat as

$$M = \frac{\text{focal length of objective}}{\text{focal lenght of eye - piece}}$$

$$\therefore \qquad M = \frac{F}{F'} = \frac{F}{3f/4} = \frac{4}{3} \cdot \frac{F}{f},$$

since the equivalent focal length F' of eye-piece is $3f/4$.

Cardinal Points : The focal length of the equivalent lens and the position of the principal foci and the unit planes may also be determined by the use of equations (2), (3) and (4).

$$F = +\frac{f_1 f_2}{f_1 + f_2 - a} = \frac{f^2}{2f - \frac{2}{3}f} = +\frac{3}{4}f,$$

$$\beta = -\frac{f_1 f_2}{f_1 + f_2 - a} = -\frac{f \cdot \frac{2}{3}f}{2f - \frac{2}{3}f} = -\frac{f}{2}$$

and $$\alpha = +\frac{f_1 a}{f_1 + f_2 - a} = -\frac{f \cdot \frac{2}{3}f}{2f - \frac{2}{3}f} = +\frac{f}{2}.$$

The positions of the principal points H_1 and H_2 and the principal foci F_1 and F_2 are marked with the aid of the above values. Since the system is in air, the nodal points coincide with the principal points. From this we can easily find that the distance of the first principal focus from the field lens of the eye-piece is given by, $F_1 A = F_1 H_1 - \alpha$

$$= \frac{3f}{4} - \frac{f}{2} = \frac{f}{4}$$

Similarly, the distance of the second principal focus from the eye-lens,

$$BF_2 = H_2F_2 - \beta = \frac{3f}{4} - \frac{f}{2} = \frac{f}{4}.$$

For Guass eye-piece which is a modification of Ramsden's eyepiece and is used in spectrometric measurements.

Huygens' Eye-piece : This eye-piece is so designed that when two converging lenses are used the condition for partial achromatisn is satisfied and the spherical aberration is *minimum.* For the combination of two thin lenses to be achromatic, the distance apart of the lenses should be half the sum of the focal lengths of the component lenses.

viz., $$a = \frac{f_1 + f_2}{2} \qquad ...(i)$$

and for minimum spherical aberration

$$a = f_1 - f_2 \qquad ...(ii)$$

Combining (*i*) and (*ii*), we get

$$f_1 - f_2 = \frac{f_1 + f_2}{2}, \text{ or } f_1 = 3f_2$$

and $$a = 3f_2 - f_2 = 2f_2.$$

The eye-piece, therefore, consists of two thin lenses so that the focal length of the field lens f_1 is 3 times the focal length of the eye lens f_2, and the distance apart twice the focal length of the eye lens. Both the lenses are plano-convex of the same kind of glass arranged with their convex faces towards the incident rays in order to reduce the various aberrations.

To trace the pencil of rays through a telescope fitted with Huygens' eye-piece, we observe that in order that the rays must emerge from the eye lens as a parallel beam, the field lens must form an image A_1B_1, at the first focal plane of the eye lens, *i.e.*, midway between the two lenses. Therefore, if AB is the position of the image formed by the object glass, then applying the relation

$$\frac{1}{v} - \frac{1}{u} = \frac{1}{f_1}$$

to the field lens, we have

$$\frac{1}{f_1} - \frac{1}{u} = \frac{1}{3f_2},$$

Since $\quad v = f_2$

and $\quad f_1 = 3f_2.$

$$\therefore \quad \frac{1}{u} = \frac{1}{f_2} - \frac{1}{3f_2} = \frac{3-1}{3f_2} = \frac{2}{3f_2}$$

or $\quad u = \dfrac{3f_2}{2}$

This shows that the image AB is *virtual* and it is on the same side of the field lens A_1B_1. Hence a parallel beam falling on the objective converges towards AB but is intercepted by the field lens and brought to focus at A_1B_1. Proceeding these rays strike the eye glass and emerge as a parallel beam.

As the rays from the objective are deviated by the field lens before the image is formed, the first focal plane of the objective is virtual and no cross-wires or a micrometer scale can be used in this eye-piece in the focal plane of the objective.

The eye-piece is, therefore, known as a negative eye-piece and is not well-adapted for telescopes. But if cross-wires are required to be used with this eyepiece, they must be placed midway between the eye and the field lenses, *i.e.*, in the position A_1B_1. Accurate measurements, however, would not be possible as the cross-wires, or the scale would be viewed by the rays which have traversed through the eye lens only, while the image would be seen by rays that have passed through both the lenses of the eye-piece and, therefore, except close to the centre on the field of view the two images would suffer from different amounts of aberration and distortion being disproportionately magnified. The eye-piece also suffers from considerable curvature of the field, the curvature being convex towards observer's eye. The field of view of this eye-piece is also inferior to the Ramsden's eye-piece. It was to avoid these defects that Ramsden's eye-piece was designed.

The equivalent focal length F' of this eye-piece is given by

$$\frac{1}{F'} = \frac{1}{f_1} + \frac{1}{f_2} - \frac{a}{f_1 f_2} = \frac{1}{3f} + \frac{1}{f} - \frac{2f}{3f^2}$$

$$= \frac{1}{3f} + \frac{1}{f} - \frac{2}{3f} = \frac{1+3-2}{3f} = \frac{2}{3f}$$

$$F' = 3f / 2.$$

It must be placed a distance f behind the eye lens to render the rays from AB parallel.

The magnifying power of a telescope fitted with Huygens' eyepiece is somewhat reduced,

$$M = \frac{F}{F'} = \frac{F}{3f/2} = \frac{2}{3} \cdot \frac{F}{f},$$

where F is the focal length of the objective.

Cardinal Points : We can also study the behaviour of a Huygens' eye-piece by determining the focal length of the equivalent lens, the positions of the principal points and first and second principal foci.

Let $f_2 = f$, say, then $f_1 = 3f_2 = 3f$.

Now $$F' = +\frac{f_1 f_2}{f_1 + f_2 - a} = +\frac{f \times 3f}{4f - 2f} = \frac{3f^2}{2f} = \frac{3}{2}f,$$

$$\beta = -\frac{f_2 a}{f_1 + f_2 - a} = -\frac{f \times 2f}{4f - 2f} = -\frac{f \times 2f}{2f} = -f,$$

$$\alpha = +\frac{f_2 a}{f_1 + f_2 - a} = +\frac{3f \times 2f}{f_1 + f_2 - a} = \frac{6f^2}{2f} = 3f.$$

Using these values the positions of the principal points H_1 and H_2 and the principal foci F_1 and F_2 *are* marked. Since the system is in air, the nodal points coincide with the principal points. The first and second principal foci are formed at a distance points of $f/2$ in front of and behind the eye lens and the principal or unit planes are at distance f behind and in front of the eye lens.

Comparison of eye-pieces

Ramsden's eye-piece	*Huygens' eye-piece*
1. The image formed by the objective is real and lies in front of the field lens at a distance $f/4$ from it. This is the position at which the cross-wires must be placed. The cross-wires and the image are equally magnified. The eye-piece is sometimes known as positive eye-piece.	1. The image formed by the objective lies behind the field lens, midway between the two lenses. No cross-wires can be used. But if cross-wires are required to be used, the cross-wires and the image will suffer from different amounts of aberration and distortion being disproportionately magnified. The eye-piece is, therefore, known as negative eye-piece.
2. It does not satisfy the condition for minimum spherical aberration.	2. It satisfies the condition for minimum spherical aberration, $a = (f_1 - f_2)$.

Contd...

Ramsden's eye-piece	*Huygens' eye-piece*
3. It does not satisfy the condition for achromatism. However, it can be made achromatic by the use of achromatic doublets.	3. It satisfies the condition for achromatism, $a = (f_1 + f_2)/2$.
4. The image formed is almost flat, *i.e.*, other aberrations are better removed.	4. The image formed is slightly convex towards the eye, *i.e.*, other aberrations such as distortion etc. are not well removed.
5. Eye-relief* is 50% greater in this case.	5. Eye-relief is too small.
6. It is used in telescopes and other optical instruments where micrometer scales are used for accurate measurements. Since it possesses a flatter field, hence scales convering the entire field can be employed.	6. It is generally used in micro-scopes and other optical instruments for qualitative purposes only.
7. It can be used as a simple microscope, since in this first principal focus lies in front of the field lens.	7. It cannot be used as a simple microscope, because the first principal focus in this case lies between the field lens and the eye lens and is virtual.

The Spectrometer

It is one of the most important optical instruments in a laboratory. It is used for the study of spectra produced by prisms and gratings, for the measurement of the dispersive powers, wavelengths of spectral lines, the refractive indices of solids and liquids etc.

(a) A collimator C, consisting of a tube of variable length with an adjustable slit S at one end and a convex lens at the other. If a sourcc of light, say, an arc lamp A is placed close to S which is at the focus of the convex lens, a parallel beam of light will emerge.

(b) A turn table R, provided with levelling screws on which a prism, a grating or any other optical device may be placed. It can be fixed at any desired height and can rotate about the central telescope T, provided generally with a Ramsden's eye-piece and cross-wires for examining the spectrum. It is also capable of rotation about a central vertical axis independent of the table. The position of the telescope is read on a graduated disc attached to it by means of two fixed verniers V_1 and V_2.

There are fixing as well as tangent screws for slow horizontal motion attached to both the table and the telescope.

Adjustments of a Spectrometer : Before using a spectrometer certain necessary adjustments have to be made.

1. *Alignment :* In the first place we must see the *alignment* of the axis of the telescope with that of the collimator so that they intersect the principal vertical axis of rotation of the telescope. This is generally done by the makers but may be checked by the student, if desired. (For this the student is advised to consult a book on Practical Physics.)
2. *Adjustment of the Telescope and Collimator for Parallel Light :*

Method 1.—The telescope alone is first directed towards a bright object, say, a lamp globe or the sky and the eye-piece is adjusted so that the cross-wires are as sharp as possible when seen through it. This fixes the position of the eye-piece relative to the cross-wires. Next pointing the telescope through an open window at some object a few hundred metres away such as a telegraph or electric post, the cross-wires and the eye-piece are racked together until a distinct image of the distant object is seen and there is no parallax of this image with that of the cross-wires.

The telescope is now turned towards the collimator with its slit illuminated with a monochromatic source of light, say, a sodium flame. The collimator slit is, then, adjusted until a distinct image of the slit falls on the cross-wires. The instrument is now adjusted so that parallel rays pass from the collimator to the telescope.

Method 2.— This method involves the use of a Gauss eye-piece and is used when a distant object is not easily available. It is a Ramsden's eye-piece with a thin plane plate of glass *G*, placed between its two components at an angle of 45° with the optical axis of the telescope. Light from a source *S* is admitted through an opening in the side of the draw tube and is reflected down the axis of the tube illuminating the cross-wires in its path. When the telescope is focused for parallel light and has its axis almost perpendicular to a plane reflecting surface placed on the prism table before the objective then an image of cross wires formed by rays reflected from the plane reflecting surface back into the objective will be at the principal focus of the objective. If this image coincides exactly with the cross-wires, the axis of the telescope is exactly normal to the reflecting surface. When the cross-wires and their images are both in sharp focus without parallax, the telescope is correctly adjusted for parallel light. For eollimator adjustment proceed as suggested in method 1.

Method 3.— Schuster's method. In the absence of a distant object or a Gauss eye-piece, the following method due to Schuster may be used :

After focusing the eye-piece on the cross-wires, the telescope and the collimator are adjusted in a straight line. The slit is illuminated with sodium light and the prism is so placed on the prism table that it has maximum illumination from the collimator. The prism table is then rotated; slowly following the image with the telescope until the position of minimum deviation is found. The telescope is then moved through about 10° so as to enable it to receive images for two positions of the prism marked *P* and *Q*, with deviation greater than the minimum. Suppose the prism is in the position

P. marked by full lines then the rays from the collimator fall on it more obliquely than for minimum deviation. The telescope is then focused' till the image is sharp and without parallax with the cross-wires. When the prism is next rotated to the position *Q*, the image will be found to be blurred. This time the adjustment of the slit be made with respect to the collimating lens till the image is in focus again. These alternatives of rotating the prism table and focusing first with telescope and then with collimator are repeated until turning the prism causes no change of focus. When this condition is achieved the rays from the slit are parallel in passing through the prism.

Note : For several measurements with the help of a spectrometer some advanced book on Practical Physics may be consulted.

The Hilger's Constant Deviation Spectrometer : In this spectrometer, the telescope and the collimator *C* are fixed at right angles to each other on a cast iron stand. A constant deviation prism *P* is mounted on the prism table which can be rotated by a fine micrometer screw to which a drum *D* is attached. The prism may be regarded as made up of two 30° prisms and one right-angled prism. For any angle of incidence i_1, the light of the wavelength which is at minimum deviation emerges at right angles to the direction of the incident beam after one internal reflection at the face *BC*. When the prism table is rotated, the spectrum; crosses the field and the drum passes under an index which indicates the wavelength of any spectral line which fails on the cross-wires.

The instrument can serve as a monochromatic illuminator. For this purpose it is necessary only to remove the eye-piece and place a writable aperture in the focal plane of the spectrum.

Spectroscope and Spectrograph : If a spectrometer is merely designed for the qualitative study of the spectra, that is, to view the various spectral lines emitted by a source, it is called a spectroscope. One convenient type, which gives a spectrum with practically no deviation is called a direct vision spectroscope.

If it is desired to have a permanent record of the spectrum of a source, the eye-piece of the telescope is removed and a photographic plate is placed in the focal plane of the objective. Such an instruments is then called a spectrogram. By noting the position of a line, or the separation between lines of different wavelengths we can find the wavelengths of the spectral lines if that of a particular line is known. These instruments are now commonly used with great success in the study of spectra especially in the ultra-violet region and the near infra-red region as such regions cannot be detected by the human eye.

For photographing these regions plates with specially prepared emulsions are used. The lenses and prisms are made of quartz as glass absorbs the light from these regions. For ultra-violet ray petrography see chapter 23.

2

Power Instruments

Resolving Power

When two objects are very close together they may appear as one and it may be difficult for the eye to see them as separate merely by magnification. The eye can see two objects as separate only if the angle subtended by them at the eye is greater than one minute which is the resolving limit of the normal eye. To separate two objects which are very close together, optical instruments such as telescopes, microscopes, prisms, gratings etc. are employed. The separation of such close objects is termed as 'resolution', and the ability of an optical instrument to produce distinctly separate images of such close objects is called its 'resolving power'.

The two neighbouring object-points are resolved by lens systems such as telescopes and microscopes, while two close spectral lines are resolved with the help of prisms and gratings. Here we, assume that the instruments' employed for the purpose are perfect, *i.e.*, they are free from all optical defects, so that according to the laws of geometrical optics the images are perfect—the image of a point object or a line is simply a point or a line. Since the laws of geometrical optics and rectilinear propagation of light are only approximate so that owing to the wave nature of light instead of the sharp images we obtain diffraction patterns the forms of which will depend upon the *shape* and *size* of the aperture.

In reality a point source, even in the absence of all lens aberrations does not give an accurate point image in the conjugate image plane, but a circular diffraction pattern consisting of a central bright disc surrounded by dark and bright concentric circular rings rapidly decreasing in intensity in the manner indicated by the curve. When two such diffraction patterns are formed close together they overlap each other and the problem of their resolution becomes more complex and difficult to deal with. The closer they are, the more difficult it becomes to distinguish them as separate.

Rayleigh's Criterion of Resolution : To obtain a definite measure of the resolving power Lord Rayleigh (1879), after a detailed study of the resultant intensity distribution in the diffraction patterns of closely situated equally bright point sources, suggested that two such patterns may be regarded as separate, that is, jut resolved if the central maximum of one coincided with the first minimum of the other, *i.e.*, when the centres of the maxima are separated by half width of the lines.

This is known as Rayleigh's criterion of limiting resolution. This is equivalent to the condition that the distance between the centres of the patterns should be equal to the radius of the central diffraction disc.

This criterion of resolution is equally applicable to the resolution of spectral lines of equal intensity. Thus two close lines in a spectrum are said to be *Just* resolved when the central maximum of the diffraction pattern of one falls upon the first minimum of the diffraction pattern of the other. The study of resultant intensity distribution in the diffraction patterns of two spectral lines having wavelengths λ and λ_1 ($= \lambda + d\lambda$) brings out clearly the physical significance of Rayleigh's criterion for resolution. The difference in wavelengths of the two the angular separation between their spectral lines is such that centres is much greater than that between the centre of either and its first minimum. There is a distinct point of zero intensity in the centre of the resultant intensity curve, so that the two spectral lines are distinctly separate.

If, however, difference in wavelengths is smaller, the central maxima of the two spectral lines are closer. The difference in wavelengths of the spectral lines is such that the central maximum of one falls exactly over the first minimum of the other and the resultant intensity curve is a double humped curve shown in thick line with a clear dip in the centre of the two central maxima.

This is easily understood if we remember that the two separate intensity curves intersect in this region at $\alpha = \pi/2$, so that the corresponding intensity in each pattern is $A_0^2 \sin^2 \alpha/\alpha^2 = A_0^2 \times 4.\pi^2 = 0.405$ times that of the central maximum. The resultant intensity is, therefore, 0.810 times (or nearly 80%) that of the principal maximum of each separate pattern. This change in intensity is easily detected by the eye and helps us in distinguishing that in the resultant diffraction pattern there are two spectral lines, *i.e.*, the lines are just resolved.

If, however, the difference in the wavelengths is so small that the principal maxima of the spectral lines come still closer and there is sufficient overlapping of the intensity curves , the resultant intensity curve shows only one maximum in the centre without any dip. Under these conditions the two spectral lines are not resolved.

Thus for the two spectral lines to be just resolved, it is essential that the principal maximum of one spectral line must fall upon the first subsidiary minimum of the other, *i.e., the least angular separation between the principal maxima of the diffraction pattern of the two spectral lines in a given diffraction order should be equal to half the angular width of either principal maximum.* The resolving power in this case is measured by $\lambda/\Delta\lambda$, where λ is the mean wavelength of a barely resolved pair and $\Delta\lambda$ is the least difference in wavelengths, $(\lambda_1 - \lambda)$ of the two spectral lines that can just be seen as separate. This ratio is called the Chromatic Resolving Power of the instrument at that wavelength.

Resolving Power of a Telescope : In using a telescope (or microscope) the main function is not merely to magnify an object, but to be able to observe finer details in it than with the unaided eye. In the case of a telescope it is to find the degree of closeness of two neighbouring stars which could be resolved.

Suppose a parallel beam of light from a distant source, incident normally on the objective of a telescope, is brought to focus at F. The supporting ring of the telescope objective or the lens itself serves as a *circular aperture.* This will give rise to a diffraction pattern consisting of a bright circular disc with centre at F surrounded alternately by a series of dark and bright rings, in the focal plane of the objective. Let the first minimum of the diffraction pattern be produced at P, then the path difference $(AP - BP) = \lambda$.

$$\text{Now,} \quad AP - BP = AE = AB \times \Delta\theta = D \times \Delta\theta$$

where AB (= D) is the diameter of the lens aperture and $\Delta\theta$ is the angular separation (= $\angle EOP$) between the central maximum and the first minimum. Since for a first minimum,

$$AP - BP = \lambda.$$

$$\therefore \qquad \lambda = D \times \Delta\theta$$

whence $$\Delta\theta = \frac{\lambda}{D}. \qquad \text{... }(i)$$

Now consider two distant point sources of light, such as two stars S and S' having an angular separation of *do*. They give rise to two circular images with centres at F and P, lying in the focal plane of the telescope objective AB, the central bright disc of each image will be surrounded by similar diffraction patterns.

The two diffraction patterns overlap each other and according to Rayleigh's criterion the two stars S and S' will be just resolved when the central maximum of the diffraction first minimum of the diffraction is equal to the angular separation between the central maximum of one and its first minimum. Hence in the limiting resolution

$$d\theta = \Delta\theta = \frac{\lambda}{D} \text{ ...[by equation } (i) \text{ above]}$$

$d\theta$ is known as *minimum angle of resolution.*

A more rigorous treatment of this condition for a circular aperture by Airy showed that 1.22 λ may be used for λ,

whence $$d\theta = 1.22 = \frac{\lambda}{D}. \qquad \text{...}(ii)$$

The expression on the right of equation (*ii*), therefore, gives the minimum angle between the two stars *resolvable* by a telescope.

Here dθ refers top the limit of resolution of the telescope, and the reciprocal of $d\theta$ (the limit of resolutions) is known as the resolving power of the telescope.

The smaller is the value of this minimum resolvable angle (δθ) *the greater is the* resolving power *of the telescope.*

The expression, $\frac{1}{d\theta} = \frac{D}{1.22\lambda}$, is, therefore, used to indicate the resolving power of the telescope.

Thus the resolving power of a telescope is defined as the reciprocal of the angular separation that the two distant point objects must have so that their images are just resolved *i.e.,* seen through die instrument as distinctly separate.

It obviously depends upon the diameter of the objective and the wavelength of the light used.

To increase the resolving power of a telescope, that is, to reduce the magnitude of the minimum resolvable angle it is necessary to increase the aperture. Moreover, large aperture would collect more light and would make even faint stars clearly visible. It is for these reasons that the diameters of telescope objectives used in astronomical observations are very large.

Also we find that since the triangles *ABE* and *POF* are equiangular and hence similar, the angle of diffraction for the first minimum,

$$-\Delta\theta = \frac{PF}{QF} = \frac{r}{f}$$

where r is the radius of the first dark ring and f the focal length of the objective.

$$\therefore \qquad \frac{r}{f} = 1.21\ \frac{\lambda}{D}, \qquad \text{...[by equation } (ii)]$$

$$\text{or} \qquad r = 1.22\ f\ \frac{\lambda}{D}$$

which shows that if fora given wavelength of incident light the greater is the diameter and smaller the focal length of the objective, the smaller would be the radius of the central bright disc called Airy's disc. The diffraction rings will, therefore, be sharper and the values of the minimum angular separation $d\theta$ for the two point-objects to be seen as distinct from each other will be smaller, thus increasing the resolving power.

The formula for the resolving power of a telescope also shows that in order to resolve stars with an angular separation of one second (4.84×10^{-6} radian), the wavelength being assumed as 5×10^{-8} cm, we would require an objective with a diameter,

$$D = \frac{1.22\,\lambda}{d\theta} = \frac{1.22 \times 5 \times 10^{-5} \text{ cm}}{4\,84 \times 10^{-8}}.$$
$$= 12.6 \text{ cm.}$$

Dawe's Rule : If the wavelength of light is taken as 5×10^{-5}cm as in the above example and the diameter of a refracting telescope objective be denoted by D cm, the minimum resolvable angle

$$d\theta = 1.22\,\frac{\lambda}{D}, \text{ radian}$$
$$= \left[\frac{1.22 \times 5 \times 10^{-5}}{D} \times \frac{180 \times 60 \times 60}{\pi}\right] \text{ seconds}$$
$$= \frac{12.4}{D} \text{ seconds.}$$

The resolvable angle of the 100 cm refracting telescope at Yerke's observatory is

$$d\theta = \frac{12.4}{100} \text{ seconds or } \frac{1''}{8}.$$

Hence we can say that the least angular separation (in seconds) resolvable by a telescope is

$$d\theta = \frac{12.4}{\text{aperture in com.}} \quad \text{[Dawe'a rule]} \qquad ...(3)$$

Relation between the Magnifying Power and the Resolving Power of a Telescope : Let us consider again the relation $r = 1.22\,F\,\lambda/D$. If F and hence the magnifying power ($M = F/f$) is increased without a corresponding increase in D, the size of the central bright disc will increase. Since every point of the object forms its own disc, the grain of the image increases.

If two neighbouring stars are being viewed, each will produce its own diffraction disc and it is obvious that unless the two discs appear separate in the primary image no amount of magnification will make them appear as such.

If, however, the two objects appear as separate in the primary image but the separation may be so small that they cannot be resolved by the eye, then the next question is what should be the minimum magnifying power of the telescope so that its theoretical resolving power is fully utilized. For this purpose the two diffraction discs which are just resolved in the focal plane of the objective must subtend at the eye an angle (say $d\theta_1$) at least equal to the limit of resolution of the eye, *viz.*, 1.22 λ/d, where d, is the diameter of the eye pupil.

Now by equation (9), diameter of the exit pupil (= d) of a telescope is given by

$$d = \frac{D}{F/f} = \frac{D}{M},$$

where D is the diameter of the entrance pupil (objective) and M the magnifying power of the telescope.

In order, therefore, that the eye should be able to just resolve the objects the magnification of the telescope should be increased by the use of an ocular of high power thus reducing the size of the exit pupil to that of the pupil of the eye, that is, d should be made equal to d, or, the magnification should be *normal**. This sets the lower limit to the magnifying power of the telescope for the practical realization of its theoretical resolving power.

∴ normal magnifying power,

$$M = \frac{D}{d_2}$$

or $$M = \frac{1.22\,\lambda / d_0}{1.22\lambda / D}$$

where $d\theta = \dfrac{1.22\lambda}{D}$ is the limit of resolution of telescope

and $d\theta_1 = \dfrac{1.22\lambda}{d_0}$ is the limit or resolution of the eye.

$$\therefore \qquad M = \frac{\text{Limit of resolution of the eye}}{\text{Limit of resolution of the telescope}}$$

$$= \frac{d\theta_1}{d\theta} \qquad \ldots (4)$$

Hence if the diameter d of the exit pupal is larger than $d.$, that of the eye pupal or the magnifying power is less than the normal, the image will case to be resolved by the eye even though they are resolved in the focal plane of the objective.

The limit of resolution of the eye according to helmets lies between 1 and 2 minutes. Let us take it as 1.5 min so that Yerke's telescope by relation (4) above must have a normal magnifying power of

$$\left(\frac{1.5 \times 60}{1/8}\right) = 720.$$

Thus to bring out the full details of the object in this telescope a magnifying power of about 720 would be required. If the magnifying power is greater than this it will serve no useful purpose since if would not be accompanied by any more resolution, *i.e.*, no fresh details of the object would be revealed and if it is less than this, the full resolving power of the instrument will not be utilized. For a given aperture, therefore, there is a limit to the useful *magnifying power* of a telescope, and any magnification in excess of that may be called *empty* magnification.

Determination of the Resolving Power of a Telescope : To determine experimentally the resolving power of a telescope, a piece of wire gauze fixed on some tissue paper is placed in front of the objective of the telescope, the tissue paper being illuminated by a sodium lamp, say. Then the distance of the gauze from the objective is adjusted till its wires just appear as separate. If D be this distance and d the separation of the two neighbouring wires as measured by a micrometer, then the ratio d/D, which is the angle subtended by the two neighbouring wires at the object glass is a measure of the practical limit of resolution.

Example : *Calculate the limit of resolution of a telescope whose object lens has a diameter of 5 m (1λ = 6 × 10⁻⁵ cm).*

Here $l = 6 \times 10^{-5}$ cm; $D = 5$ m = 500 cm.

The minimum resolvable angle

$$d\theta = \frac{1.22\,\lambda}{D}$$

$$= \frac{1.22 \times 6 \times 10^{-5}\ \text{cm}}{500\ \text{cm}}$$

$$= \frac{1.22 \times 6 \times 10^{-5}}{500}\ \text{rad.}$$

$$= \frac{1.22 \times 6 \times 10^{-7}}{5} \times \frac{180 \times 60 \times 60}{3.142}\ \text{second}$$

$$= 0.030193\ \text{seconds.}$$

Example : *Calculate the resolving power of the human eye taking $\lambda = 6 \times 10^{-3}$ cm, diameter of the pupil as 2 mm and the refractive index of vitreous humour $\mu = 1.44$.*

Here, $\lambda = 6 \times 10^{-5}$ cm ;

$d_0 = 2$ mm $= 0.2$ cm.

Effective wavelength of light inside the eye,

$$\lambda = \frac{\lambda_0}{\mu} = \frac{6 \times 10^{-5}\ \text{cm}}{1.44}$$

∴ limit of resolution of the eye,

$$d\theta = \frac{1.22\,\lambda}{d_0}$$

$$= \frac{1.22 \times 6 \times 10^{-5} \text{ cm}}{1.44 \times 0.2 \text{ cm}}$$

$$= \frac{1.22 \times 6 \times 10^{-5}}{1.44 \times 0.2} \text{ rad}$$

$$= \frac{1.22 \times 6 \times 10^{-5}}{1.44 \times 0.2} \times \frac{180 \times 60}{3.142}$$

min = 0.8736 min.

Note **:** The reader may think as to why this value of the wavelength in the vitreous humour is not taken into account in the calculation given in the text. The reason is that although the dimensions of the diffraction pattern are smaller on this account but the separation of the two images is decreased in the same ratio by refraction of the rays on entering the eye.

Example **:** *Calculate the useful magnifying power of a telescope of 2.54 cm objective assuming the limit of resolution of the eye to be equal to 2 minutes and the wavelength of light 6×10^{-5} cm.*

Here, $\lambda = 6 \times 10^{-5}$ cm; $D = 2.54$ cm.

The limit of resolution of the telescope is given by

$$d\theta = \frac{1.22\lambda}{D} = \frac{1.22 \times 6 \times 0^{-5} \text{ cm}}{2.54 \text{ cm}}$$

$= 2.881 \times 10^{-5}$ radian = 0.09906 minutes.

The angular resolution limit for the eye

$d\theta_1 = 2$ minutes.

The useful magnifying power of the telescope*

$$= \frac{\text{Limit of resolution of the eye}}{\text{Limit of resolution of the telescope}}$$

$$= \frac{d\theta_1}{d\theta} = \frac{2}{0.09906} = 20.2.$$

Example : *Calculate the least value of the angular separation of two stars which can be resolved by a telescope of 2 metres aperture. If the aperture of the human eye be 2 mm and the focal length of the eye-piece be 2.54 cm, what must be the minimum focal length of the objective if the full resolving power of the telescope is to be utilized 1 (Assume λ = 6000 Å).*

The least value of the angular separation of two stars which can be resolved by a telescope is given by $d\theta = 1.22\lambda/D$.

Here, $D = 200$ cm, $\lambda = 6000 \times 10^{-8}$ cm.

Therefore the least value of angular separation,

$$d\theta = \frac{1.22 \times 6000 \times 10^{-5}}{200}$$

radian = 36.6×10^{-5} radian.

We know that the normal magnifying power of the telescope

$$= \frac{\text{Limit of resolution of the eye}}{\text{Limit of resolution of the telescope}}$$

$$= \frac{1.22\,\lambda/d_e}{1.22\,\lambda/D} = \frac{D}{d_e},$$

where d_e is the aperture of the eye and D is the aperture of the telescope objective. This gives the magnifying power of a telescope for full utilization of its resolving power.

Here, $d_e = 2$ mm $= 0.2$ cm and $D = 200$ cm.

$$\therefore \quad \text{magnifying power} = \frac{200 \text{ cm}}{0.2 \text{ cm}} = 1000.$$

$$\text{Also magnifying power} = \frac{F}{f},$$

where F and f are the respective focal lengths of the objective and the eye-piece of the telescope.

$$\therefore \qquad 1000 = \frac{F}{2.54}$$

or $\qquad F = 2540 \text{ cm} = 25.4$ metres.

Example : *A telescope of 20 cm objective is pointed towards the moon distant* $4{\times}10^5$ *km when two objects on its surface are just resolved. Calculate (ii) the distance between them and (ii) MP. of the telescope required for the purpose,* $\lambda = 5800$ Å *and limit of resolution of eye = 1.5 min.*

Here, $\qquad \lambda = 5800 \times 10^{-8}$ cm; $D = 20$ cm.

Limit of resolution of the telescope,

$$d\theta = \frac{1.22\lambda}{D} = \frac{1.22 \times 5800 \times 10^{-8}}{20} \text{ radians.} \qquad ...(i)$$

Distance of moon from earth = 4×10^5 km. Let distance between two objects on the moon be x km.

$$\text{Then} \qquad d\theta = \frac{x.}{400{,}000} \text{ rad.} \qquad ...(ii)$$

Putting $(i) = (ii)$, we get

$$x = 1.3 \text{ km.}$$

Also M.P. of telescope

$$= \frac{\text{Limit or resolution of eye}}{\text{Limit of resolution of telescope}} = \frac{d\theta_1}{d\theta}$$

$$= \frac{1.5 \times \pi / 60 \times 180}{\frac{1.22 \times 5800 \times 10^{-9}}{20}} = 120.$$

Resolving Power of a Microscope : Here the problem is to find *the smallest linear separation which two neighbouring small objects can have and be seen as distinctly separate in the instrument.* Suppose AB is the objective of the microscope and P and Q are two small objects (assumed self luminous) placed very close to it and perpendicular to its axis. The rays starting from P, after refraction through the lens, meet at P_1 in the *same* phase giving rise to a diffraction pattern with the centre of bright central disc at P_1 surrounded by a system of alternately dark and bright rings as explained above. Similarly, the image of Q at Q_1 is a bright disc surrounded by dark and bright rings. *The points P and Q will be just resolved in the image plane if the first dark ring of the diffraction pattern of the point Q passes through* P_1 *the centre of the bright disc of P.* For this condition to be fulfilled the rays QA and QB starting from Q and diffracted by the aperture of the lens towards P_1 to produce the first minimum must differ in path by 1.22λ.

Now the path difference between these extreme rays

$$= (QB + BP_1) - (QA + AP_1)$$

$$= (QB - QA). \qquad \text{...[since } AP_1 = BP_1]$$

Draw PC and PD perpendiculars to AQ produced and QB respectively.

Now $\angle QPC = \angle QPD = \angle APO = \theta$,

which is half of the angle between the extreme rays PA and PB.

It is clear, that

$$QC = PQ \sin\theta \text{ and } QD = PQ \sin\theta$$

$\therefore$ the path difference

$$(QB - QA) = (QD + DB) - (CA - QC)$$

$$= (QD + PB) - (PA - QC)$$

$$[\because DB = PB \text{ and } CA = PA]$$

$$= QD + QC \qquad ...[\text{since } PB = PA$$

$$= PQ \sin\theta + PQ \sin\theta = 2PQ \sin\theta.$$

Equating this path difference to 1.22 λ for limiting resolution, we have

$$2PQ \sin\theta = 1.22\lambda,$$

or
$$PQ = \frac{1.22\,\lambda}{2\sin\theta}.$$

If we denote *PQ*, the linear separation between two closest objects which can be resolved by *x*, then

$$x = \frac{1.22\,\lambda}{2\sin\theta} \qquad ...(5)$$

This gives the limit of resolution of a microscope, lesser the value of *x* greater is the resolving power.

The high resolving power microscopes generally employed in the laboratories are oil-immersion type, in which both the object and the objective are immersed in a liquid of the same refractive index as glass. The liquid generally used is cedar wood oil. If μ denotes the refractive index of the liquid, λ the wavelength of light in it and λ_0 the wavelength in vacuum, then $\lambda = \lambda_0/\mu$ so that

$$x = \frac{1.22\,\lambda_0}{2\mu\sin\theta} = \frac{0.61\,\lambda_0}{(N.A.)}, \qquad ...(6)$$

where the quantity $\mu \sin\theta$ was called by Abbe the numerical aperture *(N.A.)* of the microscope. Part I. It is a quantity of great importance in connection with the resolving power of a microscope ; for the greater is its magnitude, the smaller is the value of x, that is, finer is the detail that can be observed.

Thus we find that *to diminish the distance x between the two points to be resolved by the microscope,* λ_0 *should be decreased and N.A. increased.* This is achieved in the high power microscopes as suggested above by using (*a*) oil immersion type objectives, where a transparent oil of high refractive index is used between the object and the objective, and (*b*) by using ultra-violet light for the illumination of the objects. In the latter case all optical parts must be made of quartz as glass absorbs the ultra-violet light and photography should be made use of in examining the image. Such a microscope is called ultra-microscope.

Various Microscope

The Electron Microscope : For still higher resolving powers a beam of fast moving electrons, which behaves like extremely short waves, can be employed in place of ultra-violet light. In case of electrons accelerated through a potential difference of V volts, the associated wavelength λ is given by

$$\lambda = \frac{12.27}{\sqrt{V}} \text{ Å}.$$

For $V = 10000$, the relation gives $\lambda = 0.122$ Å.

This is much shorter than the wavelength of ultra-violet light, 1000 Å. Obviously, therefore, much greater resolving powers could be obtained. Such microscopes have been constructed and are in use in big research laboratories. They are known as electron-microscopes.

Let us briefly consider the design of an electron lens. The basic idea is that of focusing a divergent electronic beam by means of electrostatic or magnetic fields in the same way as a beam of light is focused by a lens.

Electrostatic Lens : The focusing action of an *electrostatic lens.* It consists of two hollow coaxial cylinders maintained at different potentials. The potential V_1 of the first cylinder is below the potential V_2 of the second cylinder and both V_1 and V_2 are higher than the potential of the cathode which emits the electrons. A few of the equipotential lines due to the two cylinders are shown by continuous curves and the paths of a beam of electrons issuing from *A* and travelling from left to right, are indicated by the *chain lines* which converge to focus at *B*.

Magnetic Lens : The focusing action of a magnetic field on a beam of electrons confined to the axis of the magnetic coils. When electrons cross the magnetic field and their paths make an angle with the magnetic lines, they are deflected in spiral like paths, which, if properly controlled bring them to focus.

It can be shown that the focal length of an electron lens, the field strengths of either kind and the electron velocity fit into well known formulas in optics.

By a proper design of such electrostatic lenses or alternatively the magnetic lenses, the various parts of an optical microscope such as its condensing lens, objective and eye-piece, projection lens, etc. can all be duplicated electronically. A heated filament (cathode) serves as a source of electrons, which are accelerated by an electron gun and strike the object under examination.

The thinner portions of the object allow more electrons to pass through than Electron microscope, the thicker portions just like a slide in an optical lantern. Needless to say that the whole assembly is enclosed in a vacuum chamber. The final image is received on a photographic plate or on a fluorescent screen, which can be examined visually or photographed with a still further gain in magnification.

In this way an overall magnification as great as 50,000 may be obtained. Although in such an instrument, the numerical aperture is small being of the order of 0.01, yet if we apply relation (6) to 10.000 volt electrons with their associated wavelength of 0.122 Å, we find that objects as close as $0.61 \times 0.122 \times 10^{-2}$ cm or about 7.44 Å should be just separable. This theoretical value is never reached but a limit of resolution as low as 20 Å has been reached in the best type, a value sufficient for observing many individual molecules.

Phase Contrast Microscope : In the ordinary microscope the structural details of an object are made visible by the alteration of the amplitude and hence the intensity of the incident wave-front by its various parts. But there are some objects which are *perfectly* transparent and of varying thickness and refractive index so that their different parts can alter the *phase* and not the *amplitude* of the incident wave-front on its transmission through them. Hence to make such objects visible there is need for converting variations of phase into variations of intensity in the plane of the image. Prof. Zernike of Holland first gave a theory of such a method called the *phase-contrast method,* details of which are outside the scope of this book.

The practical aspect of the problem is to use a *phase plate* which is a glass plate on which is deposited an annular layer of some transparent material of such thickness that it increases the optical path by one quarter of the wavelength of the given light usually green, so that an additional phase difference $\pi/2$ is introduced in the path of the undiffracted beam.

The arrangement how this is practically achieved. There are two important additions in the ordinary microscope, an *annular diaphragm D* and the *phase plate P* ; the former is placed in the first focal contrast microscope plane of the condensing lens C and the latter in the second focal plane of the objective

O. An image of the light source is focused upon the diaphragm D by the concave mirror M. The diaphragm is so placed that parallel rays leave the condenser, falling on this transparent object on the slide S which is, therefore, illuminated by a hollow cone of parallel light. If there are no structural variations in the object and hence no diffraction this parallel beam will be focused again by the compound objective O to form an image of D on the annular layer of transparent material on the phase plate P.

But if the object is small and has some structural irregularities, it will cause diffraction of light passing through it giving a pattern like Fraunhofer pattern of a single aperture as indicated by the dotted curve below P. Since the light forming the central maximum is passing through the annular layer of the phase plate P, it will undergo a phase retardation of $\pi/2$ with respect to the diffracted light on account of this part being thicker than the rest of the glass plate by $\lambda/4$.

The interference effects in the final real image at I_g are thereby altered so as to produce marked variation in intensity at the corresponding point. It may be noted that the transparent phase plate P with an annular layer coated with a transparent substance of thickness $\lambda/4$ is placed to coincide exactly with I_1, the real image of the source formed by the objective. In the arrangement discussed the intensity is increased and the various parts of the diffracting object are clearly visible by what the microscopists call negative or bright contrast.

If the phase plate layer is made three times thicker (= $3\lambda/4$) thus producing a retardation of $3\pi/2$ in the phase of the direct light with respect to the diffracted light, the interference in the image is then destructive and the object is rendered visible by what is known as positive or dark contrast, since the intensity of the direct beam is considerably greater than that of the diffracted light. For better results the annular portion of the phase plate is made absorbing by depositing a thin metal layer on it. This film makes the intensity of the direct beam more nearly equal to the diffracted and the destructive interference will be sufficiently complete.

Relation between the Magnifying Power and the Resolving Power of a Microscope : The resolving power depends entirely upon the objective but the eye-piece plays an important part in visual observation. Even if the details of the real image formed by the objective may be well-resolved, a certain minimum magnification is necessary before the eye can see the details. A relation similar to one deduced in the case of a telescope holds good. Accordingly.

Minimum magnifying power of a microscope

$$= \frac{\textit{Limit of resolution of the eye}}{\textit{Limit of resolution of the microscope}}$$

Let us illustrate it by an example :

In an oil immersion objective, the maximum value of the numerical aperture is about 1.4. With white light of effective wavelength 5.6×10^{-5} cm such an objective can resolve small objects with a distance apart,

$$x = \frac{0.61\,\lambda}{N.A.} = \frac{0.61 \times 5.6 \times 10^{-5}\ \text{cm}}{1.4}$$

$$= 2.44 \times 10^{-5}\ \text{cm.}$$

Now, a normal eye can separate neighbouring images on the retina only if the angle subtended by them is at least 1 to 2 rain, or taking the mean, 1.5 min, or $\frac{1.5}{60} \times \frac{\pi}{180}$ radian. Therefore, the distance (y) between the images at the distance of distinct vision (= 25 cm) be such that

$$\frac{y}{25\ \text{cm}} = \frac{1.5}{60} \times \frac{\pi}{180}$$

or $y = 1.1 \times 10^{-2}$ cm approx.

Hence, the magnifying power necessary to see the resolved images of the two objects 2.44×10^{-5} cm apart must be

$$= \frac{1.1 \times 10^{-2} \text{ cm}}{2.44 \times 10^{-5} \text{ cm}} \text{ or } \quad \text{about } 450.$$

It may also be noted that for $N.A. = 1.4$,

normal $M.P.=200 \times 1.4 = 280$, [*cf.*, Eqn. (14)]

But for resolution a minimum magnifying power of 450 is needed, so the size of the exit pupil will be cut down [*cf*, Eqn. (13)] resulting in diminished illumination. So that when full *N.A.* of the objective is used, increased *M.P.* does not involve a loss in resolving power but it entails a loss in brightness. Thus if plenty of light is available, greater magnifications are frequently used because a larger image can be studied more easily even if no more details are observed. If, however, the *M.P.* less than normal is used *N.A.* is not fully utilized [*cf.*, Eqn. (14)] and hence the resolving power is reduced [*cf*, Eqn. (6)].

Example : *Calculate (a) magnifying power and (b) the resolving power of a laboratory microscope with the following data :*

N.A. as given on the objective = 0.12.

Focal length of the objective as well as eye-piece = 2.5 cm.

Distance of crosswires from objective = 15 cm.

Wavelength of light used = 6×10^{-5} cm.

(a) Magnifying power.

(i) If the final image is at infinity, using relation (10), we have

$$M.P. = \frac{L}{f_0} \times \frac{25}{f_e}.$$

Here, $L = 15 - 2.5 = 12.5$ cm.

$$\therefore \quad M.P. = \frac{12.5}{2.5} \times \frac{25}{2.5} = 5 \times 10 = 50.$$

(ii) If the final image is at the distance of distinct vision

$$M.P. = \frac{v}{u}\left(1 + \frac{D}{f_a}\right)$$

Here $v = 15$ cm, $f = 2.5$ cm,

so applying the relation

$$\frac{1}{v} + \frac{1}{u} = \frac{1}{f}, \text{ we get } u = 3 \text{ cm}.$$

$$\therefore \quad M.\ P. = \frac{15 \text{ cm}}{3 \text{ cm}}\left(1 + \frac{25}{2.5}\right) = 5 \times 11 = 55.$$

(b) The minimum resolvable distance between two self-luminous points is given by

$$x = \frac{1.22\,\lambda}{2\mu \sin\theta} = \frac{1.22\,\lambda}{2 \times N.A.}$$

$$= \frac{1.22 \times 6 \times 10^{-5} \text{ cm}}{2 \times 0.12}$$

$$= 30.5 \times 10^{-5} \text{ cm} = 0.0003 \text{ cm approx.}$$

If the points are not self-luminous as is generally the case, the minimum resolvable distance is reduced to $\lambda/2 \times N.A.$,

$$\text{whence} \quad x = \frac{6 \times 10^{-5} \text{ cm}}{2 \times 0.12} = 25 \times 10^{-5} \text{ cm} = 0.00025 \text{ cm}.$$

Example *: A microscope is used to resolve two self-luminous objects 5×16^{-5} cm apart and placed in air. Find the minimum M.P. of microscope. Limit of resolution of eye =1.5 and least distance of distinct vision = 25 cm.*

Here, the resolvable distance between two points,

$$x = \frac{1.22\,\lambda}{2\sin\theta} = 5 \times 10^{-5} \text{ cm}$$

Resolution limit of eye

$$= \frac{1.5}{60} \times \frac{\pi}{180} \text{ rad.}$$

and its linear value $= \frac{1.5}{60} \times \frac{\pi}{180} \times 25$ cm.

$$\therefore \text{M.P. of microscope} = \frac{\text{Resolution limit of eye}}{\text{Resolution limit of microscope}}$$

$$= \frac{\frac{1.5}{60} \times \frac{\pi}{180} \times 25\text{cm}}{5 \times 10^{-5} \text{ cm}} = 218.$$

Resolving Power of a Prism : When a line spectrum is formed by a prism spectroscope, an image of the slit corresponding to each line is seen. Two close lines will be resolved only if the angular separation of their images exceeds the limit of resolution of the telescope. If λ *and* $(\lambda + d\lambda)$ *are the wavelengths of two close lines which are just resolved then* $\lambda/d\lambda$ *measures the* resolving power of the spectroscopic instrument. Let *ABC* be the section of a prism.

Allow a parallel beam of monochromatic light of wavelength λ produced from a slit source *S* and a collimating lens L_1 to be incident on the face *AB* of the prism placed in the minimum deviation position. Assuming the light to be passing through the whole of the prism let *BM* be the incident wave-front and *CN* the emergent wave-front which gives rise to image I_1 in the focal plane of the telescope objective L_2. If μ be the refractive index of the material of the prism for the light of wavelength A, then since any optical path between the two successive positions *BM* and *CN* of the wave-front must be the same, we pan write

$$MA + AN = \mu BC \qquad \text{... } (i)$$

If now light of a slightly different wavelength, say $(\lambda + d\lambda)$, is allowed to be incident on the face AB along the same path and at the same angle, it will produce an emergent wave-front CN_1 and give rise to the image I_2 in the focal plane of the lens L_2.

Since with the increase in the wavelength from A to $(\lambda + d\lambda)$, the refractive index decreases from μ to $(\mu - d\mu)$, we have by the same reasoning as above

$$MA + AN_1 = (\mu - d\mu)\ BC \qquad \text{... } (ii)$$

Subtracting (*ii*) from (*i*),

$$AN - AN_1 = BC.d\mu,$$

or
$$NN_1 = BC.d\mu = td\mu,$$

where t is the thickness of the base of the prism.

If a be the width of the emergent beam CN and $d\theta$ the angle between the emergent wave-fronts, then $NN_1 = CN.d\theta = ad\theta$.

$$\therefore \qquad ad\theta = td\mu$$

or
$$d\theta = t\frac{d\mu}{a}$$

Also it has been proved that the limiting condition for resolution in case of a telescope is given by

$$d\theta = \lambda/a.$$

...[∵ a = diameter of the aperture of the telescope.]

Hence the two wavelengths λ and $(\lambda + d\lambda)$ are just resolved in a prism spectroscope if

$$\frac{\lambda}{a} = t\frac{d\mu}{a},$$

or
$$\lambda = td\mu.$$

Thus the resolving power of a prism spectroscope is given by

$$\frac{\lambda}{d\lambda} = t\frac{d\mu}{d\lambda}, \qquad \text{... (7)}$$

$\frac{d\mu}{d\lambda}$ being the chromatic dispersion of the material of the prism.

The value of $d\mu/d\lambda$ may be obtained from Cauchy's empirical formula

$$\mu = A + \frac{B}{\lambda^2}$$

where A and B are constants.

Differentiating, we get

$$\frac{d\mu}{d\lambda} = \frac{-2B}{\lambda^2}. \qquad \text{... (8)}$$

Thus the resolving power of a prism spectroscope expressed in terms of Cauchy's constant B,

$$\frac{\lambda}{d\lambda} = t\frac{d\mu}{d\lambda} = \frac{-2Bt}{\lambda^2}. \qquad \text{... (8)}$$

This shows that *the resolving power of a prism is independent of the angle of the prism but depends upon the thickness of the base, material of the prism and wavelength of light used.*

With ordinary flint glass the value of $d\mu/d\lambda$ is about 800 cm^{-1} in the yellow region so that to obtain a resolving power of 1000 just necessary to separate the two D lines of sodium a prism with a base about 1.3 cm is required.

It may also be pointed out that Rayleigh's criterion may be applied to lines of almost equal intensity. For widely unequal intensities, the difference in their wavelengths must be much greater for resolution.

Example : *The index of refraction of a certain glass is given by $\mu_C = 1.6545$, $\mu_F = 1.6635$. The wavelengths of C and F lines in the solar spectrum are 6563 and 5270 Å respectively. Calculate the length of the base of a 60° prism made of this glass which is just* *capable of resolving the sodium lines of wavelengths 5896 and 5890 Å.*

The resolving power of a prism, when it is set in the minimum deviation position for light of wavelength λ, is given by

$$\frac{\lambda}{d\lambda} = t\frac{d\mu}{d\lambda},$$

where $d\lambda$ is the smallest wavelength difference and t is the length of the base of the prism.

Here, $d\lambda = 5895 \text{ Å} - 5890 \text{ Å} = 6\text{Å};$

$$\lambda = \frac{5891 \text{ Å} + 5896 \text{ Å}}{2} = 5893 \text{ Å},$$

$$\frac{d\mu}{d\lambda} = \frac{\mu_F - \mu_C}{\lambda_C - \lambda_F} = \frac{1.6635 - 1.6545}{(6563 - 5270) \times 10^{-8} \text{ cm}}$$

$$= \frac{0.0090}{1293 \times 10^{-8}} \text{ cm}^{-1}.$$

Substituting these values in the above expression, we have

$$\frac{5893 \text{ Å}}{6 \text{ Å}} = t\frac{0.0090 \text{ cm}^{-1}}{1293 \times 10^{-8}}$$

or $$t = \frac{5893}{6} \times \frac{1293 \times 10^{-8} \text{ cm}}{0.009} = \frac{5893 \times 1223}{6 \times 9 \times 10^{5}} \text{ cm}$$

$$= 1.41 \text{ cm}.$$

Resolving Power of a Grating : Just like a prism spectroscope 'the resolving power of a diffraction grating' is defined as its ability to show two neighbouring lines in a spectrum as separate. It is also measured by the ratio $\lambda/d\lambda$, where λ is the wavelength of a spectral line and $d\lambda$ is the least difference in the wavelengths of two neighbouring spectral lines which can just, be resolved. A beam of light consisting of wavelengths λ and $(\lambda + d\lambda)$ is incident normally on the surface of a plane transmission grating *AB*, the light of each wavelength would form a separate diffraction pattern of the spectroscopic slit.

According to Rayleigh's criterion they can be just distinguished as separate in any order of the spectrum, if the principal maximum of the one coincides with the first subsidiary minimum of the other. The angular deviation θ for the *n*th order principal maximum for she wavelength A along *BC* is given by

$$(a + b) \sin \theta = n\lambda,$$

where $(a + b)$ is the grating constant.

If for a small change $\Delta\theta$ in the angle of diffraction the path difference between the extreme rays changes by one wavelength, then in such a case the path difference between any pair of corresponding spaces in the two halves of the grating will differ by half a wavelength. These two halves of the issuing wave-front will thus mutually cancel out each other's effect by interference. This gives the condition for the first *subsidiary* minima on either side of the *n*th order maximum.

If the total number of lines in the grating is *N*, then the path difference between any pair of corresponding points in the two adjoining slits changes by λ/N and the condition for the subsidiary minimum in a direction making an angle $(\theta + \Delta\theta)$ with the grating normal is given by

$$(a + b) \sin (\theta + \Delta\theta) = n\lambda/N \quad \text{... } (i)$$

$$\therefore \quad \Delta\theta = \frac{\lambda}{N(a+b)\cos\theta}$$

Now, if $d\theta$ be the separation between wavelengths λ and $(\lambda + d\lambda)$ for the *n*th order principal maximum, then

$$(a + b) \sin(\theta + d\theta) = n(\lambda + d\lambda), \qquad \text{... } (ii)$$

$$\therefore \qquad d\theta = \frac{nd\lambda}{(a+b)\cos\theta}.$$

For resolution $d\theta$ must be greater than $\Delta\theta$ or in the *limiting* resolution,

$$d\theta = \Delta\theta,$$

or
$$\frac{nd\lambda}{(a+b)\cos\theta} = \frac{\lambda}{N(a+b)\cos\theta}$$

or
$$nd\lambda = \frac{\lambda}{N}$$

$$\therefore \textit{R.P.}, \qquad \frac{\lambda}{d\lambda} = \text{N} \times \text{n}. \qquad \text{... } (9)$$

It is quite obvious that the resolving power is *independent* of the grating constant but *increases* with (*i*) the order n of the spectrum and (*ii*) the total number of lines N in the effective aperture of the grating.

Dispersive Power and Resolving Power of a Grating : Let us distinguish between the dispersive power and the resolving power in the case of a diffraction grating.

(1) The dispersive power has already been defined, as measuring the ratio of the change in the angle of diffraction to the corresponding change in wavelength, that is, if d and $(\theta + d\theta)$ be the angles of diffraction in a particular order of the spectrum for wavelengths λ and $(\lambda + d\lambda)$, then the dispersive power is given by $d\theta/d\lambda$.

Now, $\qquad (a + b) \sin\theta = n\lambda,$

Differentiating, we have

$$(a+b)\cos\theta \frac{d\theta}{d\lambda} = \text{n}.$$

$$\therefore \qquad \frac{d\theta}{d\lambda} = \frac{n}{(a+b)\cos\theta} \qquad \text{... (10)}$$

viz., the dispersive power *increases* with the *order* of the spectrum and varies *inversely* as the grating element but it is *independent* of the grating width.

The resolving power $\lambda/d\lambda$ on the other hand, depends upon the separation of the central maximum of a wavelength from its first minimum. If this separation is small, the resolving power would be greater as the overlapping of a neighbouring wavelength would be reduced. Since this separation is given by

$$\Delta\theta = \frac{\lambda}{N(a+b)\cos\theta} \qquad \text{... (11)}$$

we infer that larger the total number of lines N in the grating with the given grating constant, the smaller would be $\Delta\theta$ and greater, therefore, the resolving power, that is, resolving power *increases* with grating width. The reader should not, therefore, confuse the dispersive power with the resolving power, which as just explained denotes the degree of closeness the two wavelengths can have and still can be distinguished as separate.

(2) The distinction between the two is further clearly brought out. The images of two wavelengths λ and $(\lambda + d\lambda)$, which have equal angular separation and hence the same dispersion. They are taken with the help of gratings having the same grating constant but different widths. In *(a)* the width of the grating is large and the images are sharper and narrower, whereas in *(b)* the width of the grating is small and so the images are much wider and would merge into each other much sooner than in *(a)* and therefore, resolving power in *(b)* would be much less.

(3) We have seen above that the resolving power is equal to Nn, where n is the order of the spectrum and N the total number of lines, while the dispersive power depends upon the number of lines per cm. So two gratings with the same grating elements will give the same dispersive power but one with larger number of lines, that is, greater width will give greater resolving power and the same grating partly covered, *i.e.*, with smaller width would give lesser resolving power. This means that images would be narrow and sharp if N is large as in (*a*) above and wide and overlapping if N is small as in (*b*) above.

(4) A useful relation between the resolving power and the dispersive power may be obtained by combining equations (9) and (10). Substituting the value of n from equation (10) in equation (9),

we have $$\frac{\lambda}{d\lambda} = N\,(a + b)\cos\theta\,\frac{d\theta}{d\lambda},$$

where $N\,(a+b)\cos\theta$ is the total effective aperture.

$\therefore$ resolving power = dispersive power × total effective aperture. ... (12)

Echelon Grating

For the resolution of higher order spectra which are very faint, a high resolving power grating is needed. Since the resolving power of a grating is equal to the product of the total number of lines N and the order of the spectra n, *i.e.*, $R.P. = N \times n$, obviously to increase the resolving power we have to increase N and $n \times A$ plane ruled grating can be employed upto the second order and a concave grating upto the fifth or the sixth order only, because overlapping of various order spectra and loss of intensity become very serious problems.

Also due to mechanical difficulties it is not possible to rule more than 10,000 lines to a cm. More so the rulings deviate from the ideal of equal spacings which error may be perfectly random, continuously increasing in one direction and even periodic across the surface of the grating and gives rise to ghosts or false lines accompanying the principal maxima of the ideal grating.

This led Michelson to devise a grating of a high resolving power with a fairly small value of N (say 20 to 40), but by increasing n, the order of the spectrum. *The order of the spectrum is equal to the path difference measured in wavelengths between the diffracted rays from the two corresponding points of a grating.* This path difference can be increased by introducing a glass plate in the path of one of the interfering beams.

This was done with the help of an echelon grating which consists of a number of rectangular plates of quartz or glass of *exactly* equal thickness d, all cut from a *single* plane parallel plate of *optically worked* glass and arranged in the form of a flight of steps with a constant off-set a, which is usually 1 mm. The echelon grating can be a transmission type and also a reflection type in which case its surfaces are made highly reflecting by a thin metallic coating.

Theory of the Echelon Grating : Let a parallel beam of monochromatic light of wavelength λ striking a transmission grating normally be diffracted through a small angle θ.

The path difference between the two successive rays from the corresponding points A and C after transmission

$$= \mu BC - AM$$

$$= \mu BC - (BN - BL)$$

$$= \mu BC - BC \cos\theta + AB \sin\theta$$

$$= \mu d - d \cos\theta + a \sin\theta = n\lambda$$

...for the nth order spectrum.

If θ is small, cos θ = 1 and sin θ = 0, so that

$$\mu d - d + d\theta = n\lambda,$$

or $$(\mu - 1)d + d\theta = n\lambda. \qquad \text{... (13)}$$

For θ = 0, $(\mu - 1)\, d = n\lambda$

or $$n = \frac{(\mu - 1)d}{\lambda}. \qquad \text{... (13a)}$$

If further μ = 1.5, $n = \dfrac{d}{2\lambda}$ ... (14)

Taking d = 1 cm and $\lambda = 5 \times 10^{-5}$ cm

$$n = \frac{(1.5 - 1)\text{ cm}}{5 \times 10^{-5}\text{ cm}} = 10{,}000.$$

The value of n is enormously high as compared with highest order of spectrum produced by a plane grating.

Dispersion and Resolving Power : The dispersive power ($D = d\theta/d\lambda$) of an echelon grating may be obtained by differentiating equation (13), whence

$$D = \frac{d\theta}{d\lambda} = \frac{n}{a} - \frac{d}{a} \cdot \frac{d\mu}{d\lambda}. \qquad \text{... (15)}$$

In order to express D in terms of measurable quantities, it is desirable to eliminate n between equations (13a) and (15).

Substituting the approx. value of

$$n = \frac{(\mu - 1)d}{\lambda} \qquad \text{... [from equation (13a)]}$$

in equation (15), we obtain

$$D = \frac{d}{d\lambda}\left[(\mu - 1) - \lambda\frac{d\mu}{d\lambda}\right] \qquad \text{... (16)}$$

Now the resolving power is given by the product of the total aperture A and dispersive power D *cf.* equation (12).

$$\therefore \quad R.P. = A \times D$$

$$= Na - \frac{d}{d\lambda}\left[(\mu - 1) - \lambda\frac{d\mu}{d\lambda}\right] \qquad \text{... (17)}$$

since the total aperture A is the product of the number of plates λ and the width a of each step.

For small angles, $(\mu - 1) = n\lambda/d$. *cf.* equation (13*a*).

Putting this value of $(\mu - 1)$ in equation (17), we obtain

$$R.P. = Nn - Nd\frac{d\mu}{d\lambda} \qquad \text{... (18)}$$

The second term on the right hand side is small compared to first so that to a high degree of approximation

$$R.P = Nn,$$

the same as for the diffraction grating.

If the number of plates $N = 25$ approx.

$$R.P. = Nn = 25 \times 10{,}000 = 25 \times 10^4.$$

Or more accurately, the resolving power,

$$R.P. = Nn - Nd\frac{d\mu}{d\lambda}$$

$$= 25 \times 10{,}000 - 25 \times 1 \text{ cm} \times (-800) \text{ cm}^{-1}$$

$$\text{...} \left[\text{assuming } \frac{d\mu}{d\lambda} = -800 \text{ cm}^{-1}\right]$$

$$= 25 \times 10^4 + 2 \times 10^4$$

$$= 27 \times 10^4$$

On working out in the region 6000 Å, we find that two wavelengths separated by .024 Å will be nearly resolved for the order of spectrum observed along the normal.

Similarly, for a reflection echelon grating the path difference between the successive diffracted rays from the corresponding points A and C after reflection is given by

$$BC + AM = BC + (BN - BL)$$
$$= d + d \cos \theta - \sin \theta$$
$$= d(1 + \cos \theta) - a \sin \theta$$
$$= n\lambda.$$

For $\theta = 0,\ 2d = n\lambda,$

or
$$n = \frac{2d}{\lambda}. \quad \text{... (19)}$$

Comparing equation (19) with equation (14) we find that the *resolving power of a reflection grating is four times that of a transmission grating of same specification* Thus, *R.P.* in this case is far greater than that obtained with any other grating. Further, because it concentrates all the light in one or two spectra it is very useful for the study of hyperfine structure such as the split lines in the Zeeman effect.

Example : *A grating has 1000 lines ruled on it What is the difference between the two wavelengths that just appear separated in the first order spectrum in the region of wavelength* $\lambda = 6 \times 10^{-5}$ *cm?*

Suppose λ and $(\lambda + d\lambda)$ are the two wavelengths that can just be resolved by the grating in the first order spectrum.

Now in a grating the resolving power is measured by

$$\frac{\lambda}{d\lambda} = n \times N,$$

where N is the *total* number of lines ruled on the grating and n is the order of the spectrum.

Here, $N = 1000;\ n = 1;\ \lambda = 6 \times 10^{-5}$ cm

$$d\lambda = \frac{\lambda}{n \times N} = \frac{6 \times 10^{-5} \text{ cm}}{1 \times 1000}$$

$$= 6 \times 10^{-8} \text{ cm} = 6\text{Å}.$$

Thus under the given conditions two lines differing in wave length by 6 Å will just appear separated in the first order spectrum.

Example : *Using a diffraction grating having 1000 lines ruled upon its surface it is found that the two wavelengths 5896 Å and 5890 Å are just resolved in the first order. Explain what will happen if half the grating is covered so that no light pastes through this half.*

The expression for the resolving power with these wavelengths

$$\frac{\lambda}{d\lambda} = \frac{5893\,\mathring{A}}{(5896-5890)\,\mathring{A}} = \frac{5893}{6} = 982.$$

Since it is almost equal to $nN = 1 \times 1000$, the two wavelengths 5890 and 5896 are just resolved in the first order. This means that the angular separation $d\theta$ between the spectral lines is just equal to $\Delta\theta$, the separation between the first principal maximum and the first secondary minimum close to it.

If one half of the grating is covered, the angular separation $d\theta$ of the two spectral lines would still remain the same as with the full grating, being dependent on the grating element only, but each image would expand because the separation $\Delta\theta$ between the first principal maximum and the first minimum close to it, being equal to $\lambda/N(a + b)\cos\theta$, would be doubled as N is halved. This will result in the overlapping of the spectral lines as they were in the limiting resolution in the first case and will, therefore, no longer be seen as separate.

Example : *Calculate the number of lines per cm on a plane transmission grating which when used on a spectrometer with telescope and collimator of aperture 2 cm will just resolve the two D-lines in the sodium spectrum in the second order D_1 = 5896 Å and D_2 =5890 Å.*

The angular separation between the sodium *D*-lines in the *n*th order is obtained by differentiating the grating equation

$$g \sin\theta = n\lambda,$$

where g is the grating element.

This gives $g \cos\theta \, d\theta = n d\lambda$,

or
$$d\theta = \frac{n d\lambda}{g\cos\theta} = \frac{n d\lambda}{g\sqrt{1-\sin^2\theta}}$$

$$= \frac{n d\lambda}{g\sqrt{1-\frac{n^2\lambda^2}{g^2}}} = \frac{n d\lambda}{g} \text{ approx.}$$

the terms $n^2\lambda^2/g^2$, being much smaller than 1, are neglected.

From the given data, $n = 2$; $d\lambda = 6 \times 10^{-8}$ cm, we have

$$d\theta = \frac{2\times 6\times 10^{-8}\text{ cm}}{g}$$
$$= 2 \times 6 \times 10^{-8}\text{ cm} \times x,$$

where $x\left(=\frac{1}{g}\right)$ is the number of lines per cm.

For a telescope the minimum resolvable angle is given by

$$\Delta\theta = \frac{1.22\,\lambda}{D}.$$

Here, the mean wavelength of the sodium line,

$$\lambda = 5893 \times 10^{-8}\text{ cm}, D = 2\text{ cm}.$$

$$\therefore \quad \Delta\theta = \frac{1.22\times 5893\times 10^{-8}\text{ cm}}{2\text{ cm}}.$$

For just resolution of *D*-lines of sodium in the second order.

$$d\theta = \Delta\theta,$$

Hence, we get

$$2 \times 6 \times 10^{-8} \text{ cm} = \frac{1.22 \times 5893 \times 10^{-8} \text{cm}}{2 \text{ cm}},$$

whence the number of lines per cm,

$$x = \frac{1.22 \times 5893 \times 10^{-8}}{2 \times 6 \times 10^{-8} \times 2} = 299.$$

Resolving Power of Fabry and Perot Interferometer : The fringe pattern in this case is without secondary maxima and minima. So Rayleigh's criterion for resolution of spectral lines as such will not be applicable. It is reasonable, however, to assume that two spectral lines of equal intensity may be said to be just resolved if the resultant intensity due to overlap at a point *C* midway between the images will be $8/\pi^2$ or about 0.81 times. The maximum of the either, giving a clear dip as at *D*. The intensity of each fringe then at the point *C* is 0.405 times the peak intensity of either

$$\left(\frac{I}{I_0} = \frac{\sin^2 \alpha}{\alpha^2} = \frac{\sin^2 \frac{\pi}{2}}{\left(\frac{\pi}{2}\right)^2} = \frac{4}{\pi^2} \right).$$

Then for resolution, $\dfrac{I}{I_{\max}} = 0.405.$

Now in the case of this interferometer

$$\frac{I}{I_{\max}} = \frac{1}{1 + \frac{4R}{(1-R)^2} \sin^2\left(\frac{\delta}{2}\right)}$$

Since $$\frac{a^2 T^2}{(1-R)^2} = I_{max} \qquad \text{... } cf. \text{ Eqn. (9)}$$

$\therefore$ $$0.405 = \frac{1}{1 + \frac{4R}{(1-R)^2} \sin^2\left(\frac{\delta}{2}\right)}$$

or simplifying, $\sin^2\left(\frac{\delta}{2}\right) = \frac{(1-0.405)(1-R)^2}{1.62\,R}$.

The solution of this gives the value of δ for which the intensity has dropped to 0 405 of the maximum,

$$\text{whence} \qquad \delta = 2\sin^{-1}\left[\frac{(1-0.405)(1-R)^2}{1.62\,R}\right]^{\frac{1}{2}}$$

$$= 2\sin^{-1}\left[\frac{0.367\,(1-R)^2}{1.62\,R}\right]^{\frac{1}{2}} \qquad \ldots (20)$$

Indicates, in two adjacent orders, two fringes which are just resolved and for which the wavelengths are λ and λ + $d\lambda$ respectively. Now δ is the phase difference corresponding to a fractional change in fringe order represented by *HK*. Total phase change between two maxima, *i.e.*, for *HL* it will be 25, that is,

$$4\sin^{-1}\left[\frac{0.367\,(1-R)^2}{R}\right]^{\frac{1}{2}}$$

Since a phase change of 2π is equivalent to unit order separation, the phase change of 2δ corresponds to an order change

$$dn_0 = \frac{2\delta}{2\pi}$$

$$\text{whence} \qquad dn_0 = \frac{\delta}{\pi} = \frac{2}{\pi}\sin^{-1}\left[\frac{0.367\,(1-R)^2}{R}\right]^{\frac{1}{2}} \qquad \ldots (21)$$

Since $n\lambda = 2d\cos\theta$ by equation (8), the intensity at the centre of the pattern will depend upon the difference of path for $\theta = 0$. Let n_0 be the order of interference at the centre.

For this case the equation becomes

$$n_0\lambda = 2\delta$$

or
$$n_0 = \frac{2d}{\lambda} \qquad \dots (22)$$

where n_0 is used to indicate the order of interference at the centre of the ring system. The change in the order dn_0 corresponding to the change in wavelength $d\lambda$ is obtained by differentiating (22), whence, we get

$$n_0 d\lambda + \lambda dn_0 = 0$$

or
$$\frac{\lambda}{d\lambda} = -\frac{n_0}{dn_0} \qquad \dots (23)$$

i.e., the resolving power $\left(\frac{\lambda}{d\lambda}\right)$ defined as the ratio of the wavelength to the smallest change in wavelength which can be detected, is equal to the order of interference at the centre of the ring system divided by the smallest difference of order (dn_0) which can be recognised. Then within the limits already imposed, inserting the value of dn_0 from Eqn. (21) in Eqn. (23), we have

$$R.P. = \frac{\lambda}{d\lambda} = -\frac{n_0}{dn_0}$$

$$= -\frac{\pi n_0}{2\sin^{-1}\left[\frac{0.367\,(1-R)^2}{R}\right]^{\frac{1}{2}}} \qquad \dots (24)$$

Equation (24) gives the theoretical resolving power of the Fabry-Perot interferometer. The negative sign before the expression for resolving power means simply that a positive increase in wavelength corresponds to a negative change of order, dn_0. For a wavelength of 5000 Å and mirror separation of 1 cm, it follows from equation (22) that n_0 is 40,000. The resolving power may then be calculated from equation (24) for different reflection coefficients.

This interferometer can be used for

(i) measurement of l,

(ii) determination of dl,

(iii) hyperfine structure analysis, and

(iv) evaluation of length of metre.

Example : *Calculate whether the Fabry-Perot elaton can resolve two lines at 5000 Å which are 0.01 Å apart. Take reflecting power R = 80% and plate separation d = 1 cm.*

Here λ = 5000 Å,d/λ = 0.01 Å, d = 1 cm, R = 0.8.

Minimum *R.P.* required to resolve the lines with $d\lambda$ = 0.01 Å,

$$\frac{\lambda}{d\lambda} = \frac{5000}{0.01} = 500{,}000.$$

R.P. of elaton,

$$\frac{\lambda}{d\lambda} = \frac{\pi n_0}{2\sin^{-1}\left[\frac{0.367\,(1-R)^2}{R}\right]^{\frac{1}{2}}}$$

$$= \frac{\pi}{2\sin^{-1}\left[\frac{0.367\,(1-R)^2}{R}\right]^{\frac{1}{2}}} \times \frac{2d}{\lambda}$$

$$= \frac{3.142}{2\sin^{-1}\left[\frac{0.367\,(1-0.8)^2}{0.8}\right]^{\frac{1}{2}}} \times \frac{2 \times 1}{5 \times 10^{-5}}$$

= **532,000.**

$\therefore$ elaton would resolve the given lines.

3

The Aberrations

Lens Aberrations

The simple theory of the formation of images by lenses discussed in the previous chapters was based on the assumptions that the lens aperture is small and that object lay in the neighbourhood of the axis, so that the angles of incidence and the slopes of the rays were also small. These limitations, however, do not meet our practical requirement, in the use of optical instruments such as telescopes and compound microscopes.

These instruments are designed to produce bright and magnified images and for that purpose their objectives must have very wide apertures so as to collect as large an amount of light as possible and also give greater resolving powers. We shall learn later in the chapter on diffraction that even if the lens were perfect, a sharp and well defined image will be obtained only by a beam having a wide cross-section. Moreover, in the case of telescopes, microscopes and camera lenses we wish to obtain images of points much farther from the axis. So in actual practice we cannot restrict ourselves only to rays very close to the axis, *i.e., paraxial rays,* but of necessity have to make use of wide-angled pencils of rays. Hence it is necessary to study how a lens behaves when (*a*) the aperture of the lens is large and (*b*) objects cover a wide field of view, that is oblique and large angled pencils of light are incident on it.

We find that in such cases the behaviour of a lens deviates from that given by simple theory. These deviations were first investigated by Seidel in 1855 and are called Seidel or monochromatic aberrations of a lens because they exist even if the refractive index of the material is constant, that is, monochromatic light is used.

Besides these defects in a lens, there are others due to its *prismatic* action (dispersion), when the light consists of various colours. Here μ is not constant. This later type of defect is known as chromatic aberration.

Monochromatic Aberrations

There are five monochromatic aberrations of a lens : (1) spherical aberration; (2) coma; (3) astigmatism; (4) curvature of the field; and (5) distortion.

As said above these aberrations are due to the rays which do not confine to the paraxial region. Now by Maclaurin's theorem the expression for sine of an angle gives

$$\sin\theta = \theta - \frac{\theta^3}{3!} + \frac{\theta^5}{5!} - \frac{\theta^7}{7!} + \ldots$$

In the *First Order Theory or Gauss Theory* upon which the simple formulae of the preceding chapters were based we took account of the first term in this expression only. In general for oblique or wide angled pencils there are departures from first order theory resulting in aberrations which may be calculated by retaining the second or the third terms of the above series. Since the second term contains the angle raised to third power, its inclusion leads to the third order theory and so on. According to this theory the aberration for any ray is represented by five correction terms called the Seidel sums. If the lens were to be free from all aberrations so as to behave in accordance with first order theory, each of these five terms must be zero. This is, however, a difficult condition to satisfy and when one or more of these terms do not vanish, it would indicate that some form of aberration is present. The five Seidel conditions therefore gave the five monochromatic aberrations stated above. We shall discuss them one by one.

Spherical Abberation : Consider a wide parallel beam of light falling on a bi-convex lens in a direction parallel to the principal axis. Rays lying close to the axis (paraxial rays) after refraction are brought to focus to a single point F_a, while the rays through the peripheral portion of the lens, the so called marginal rays converge to another point F_m on the axis nearer the lens than F_a. This is due to the fact that rays from any given zone of the lens have a focal length which varies with the radius of the zone so that the lens does not give a point image of a distant point source on the axis. *This failure of the central and the marginal rays to converge to a single point after refraction through a lens is called* spherical aberration.

The distance F_mF_a is taken as a measure of this *axial, longitudinal* or *central* spherical aberration for a given zone of the lens. It is said to be *positive* when marginal rays intersect on that side of the paraxial focal point from which the light is incident. Reasoning as above we can show that for a divergent lens the spherical aberration is *negative.* The best possible position of the image is at a point between F_m and F_a, where the section *(XY)* of the refracted beam by a transverse plane will have least area — the circle of least confusion. The radius of this circle is the measure of the lateral spherical aberration.

To demonstrate spherical aberration take a plano-convex lens of about 15 cm aperture (such a lens is easily available in a good laboratory). Cover its plane side with a circular card-board disc with say, seven equally spaced holes along a diameter. Illuminate it by parallel light coming from an arc lamp. The refracted beams emerging out of the holes may be received on a screen which can be moved forwards and backwards. The path of the beams can be made visible by means of smoke produced by burning *dhoop.* You will find that central rays come to focus farther off from the lens than the marginal rays.

From the above discussion it is evident that spherical aberration will be present in the case of a point object anywhere on the axis but its amount varies with the position of the object point. We shall now find out how the axial spherical aberration for any fixed point depends upon the radius of the zone of the lens through which the rays pass, the focal length, the radii of curvature of its surfaces and its refractive index.

Aberration of a Ray Refracted at a Spherical Surface : Let a concave spherical surface AMB of radius r separate a medium of refractive index μ from air and C be its centre of curvature. Let a ray PA starting from a point object P on its axis be incident at A, then on refraction it proceeds along AR and appears to come from Q', a point on the axis. Let θ and ϕ be the angles of incidence and refraction at A and $h = AN$ perpendicular from A on MP. By Snell's law :

$$\mu = \frac{\sin\theta}{\sin\phi} = \frac{\sin\theta}{\sin\gamma} \times \frac{\sin\theta}{\sin\phi} = \frac{CP}{AP} \times \frac{AQ'}{CQ'}$$

$$\mu AP \times CQ' = CP \times AQ'.$$

Now, $$AP^2 = AC^2 + CP^2 - 2AC \times CP \cos ACP$$

$$= (-r)^2 + (-u + r)^2 - 2 \times (-r)(r - u) \times (-\cos\gamma)$$

$$= 2r^2 + u^2 - 2ur - 2\left(r^2 - ur\right)\left(1 - \frac{h^2}{2r^2}\right),$$

$$\because \qquad \cos\gamma = \sqrt{1-\sin^2\gamma} = \sqrt{1-(h/r)^2} = 1 - \frac{h^2}{2r^2} \text{ approx.}$$

Simplifying $AP^2 = u^2 - \frac{h^2}{r}(u-r)$

$$\therefore \qquad AP = \sqrt{u^2 - \frac{h^2}{r}(u-r)}$$

$$= u\left[1 - \left(\frac{1}{r} - \frac{1}{u}\right)\frac{h^2}{u}\right]^{\frac{1}{2}}$$

$$= u\left[1 - \left(\frac{1}{r} - \frac{1}{u}\right)\frac{h^2}{2u}\right] \text{... by binomial expan.}$$

Similarly, $AQ' = v'\left[1 - \left(\frac{1}{r} - \frac{1}{v'}\right)\frac{h^2}{2v'}\right]$

The equation $\mu AP \times CP' = CP \times AQ'$ becomes

$$\mu\left(r - v'\right)u\left[1 - \left(\frac{1}{r} - \frac{1}{u}\right)\frac{h^2}{2u}\right] = \left(r-u\right)v'\left[1 - \left(\frac{1}{r} - \frac{1}{v'}\right)\frac{h^2}{2v'}\right]$$

Dividing throughout by $uv'r$, we have

$$\mu\left(\frac{1}{v'} - \frac{1}{r}\right)\left[1 - \left(\frac{1}{r} - \frac{1}{u}\right)\frac{h^2}{2u}\right] = \left(\frac{1}{u} - \frac{1}{r}\right)\left[1 - \left(\frac{1}{r} - \frac{1}{v'}\right)\frac{h^2}{2v'}\right]$$

which, on re-arrangement, gives

$$\frac{\mu}{v'} - \frac{1}{u} = \frac{\mu - 1}{r} + \left(\frac{1}{r} - \frac{1}{u}\right)\left(\frac{1}{r} - \frac{1}{v'}\right)\frac{h^2}{2v'}$$

$$-\left(\frac{1}{r}-\frac{1}{u}\right)\left(\frac{1}{r}-\frac{1}{v'}\right)\frac{h^2\mu}{2u}$$

$$=\frac{\mu-1}{r}+\left(\frac{1}{r}-\frac{1}{v'}\right)\left(\frac{1}{r}-\frac{1}{u}\right)\left(\frac{1}{v'}-\frac{\mu}{u}\right)\frac{h^2}{2} \quad \text{... } (i)$$

Since h^2 is small, the approximate value of v' is given by v in the equation

$$\frac{\mu}{v}-\frac{1}{u}=\frac{\mu-1}{r}, \quad \text{... } (ii)$$

where v is the distance of the image Q for paraxial rays, so that we can write

$$\frac{\mu}{v'}-\frac{1}{u}=\frac{\mu-1}{r},$$

or

$$\frac{1}{v'}=\frac{1}{\mu u}+\frac{\mu-1}{\mu r}.$$

Using this value in the expansion of the terms containing h^2 in (*i*) above, re-arranging and simplifying, we have

$$\frac{\mu}{v'}-\frac{1}{u}=\frac{\mu-1}{r}+\left[\frac{1}{r}-\left(\frac{1}{\mu u}+\frac{\mu-1}{\mu r}\right)\right]$$

$$\times\left(\frac{1}{r}-\frac{1}{u}\right)\left[\frac{\mu-1}{\mu r}-\frac{\mu}{u}+\frac{1}{\mu u}\right]\frac{h^2}{2}$$

$$=\frac{\mu-1}{r}+\frac{1}{\mu}\left(\frac{1}{r}-\frac{1}{u}\right)\times\left(\frac{1}{r}-\frac{1}{u}\right)$$

$$\left[\frac{\mu-1}{r}-\frac{\mu^2-1}{\mu r}\right]\frac{h^2}{2}$$

$$=\frac{\mu-1}{r}+\left(\frac{\mu-1}{\mu^2}\right)\left(\frac{1}{r}-\frac{1}{u}\right)^2\left[\frac{1}{r}-\frac{\mu+1}{u}\right]\frac{h^2}{2} \quad \text{.... } (iii)$$

This equation determines the position of Q'. The aberration of the ray is $QQ' = u' - v$ so that substracting (*ii*) from (*iii*), we get

$$\left(\frac{u}{v'} - \frac{1}{u}\right) - \left(\frac{\mu}{v} - \frac{1}{u}\right) = \frac{\mu - 1}{\mu^2}\left(\frac{1}{r} - \frac{1}{u}\right)^2 \left(\frac{1}{r} - \frac{\mu + 1}{u}\right)\frac{h^2}{2}$$

$$\therefore \qquad v' - v = \frac{\mu - 1}{\mu^2}\left(\frac{1}{r} - \frac{1}{u}\right)^2 \left(\frac{1}{r} - \frac{\mu + 1}{u}\right)\frac{h^2 v^2}{2}.$$

$$[\because v' \to v] \quad \ldots (1)$$

Spherical Aberration in a Thin Lens : Let us consider a thin converging meniscus of refractive index μ and radii of curvature r_1 and r_2 and P a point object on the axis. A ray PA incident on the first surface is refracted along AB and then along BC at the second surface. Then for refraction at the first surface by equation (*iii*) above, we have

$$\frac{\mu}{v'} - \frac{1}{u} = \frac{\mu - 1}{r_1} + \frac{\mu - 1}{\mu^2}\left(\frac{1}{r_1} - \frac{1}{u}\right)^2 \left(\frac{1}{r_1} - \frac{\mu + 1}{u}\right)\frac{h^2}{2}, \quad \ldots (i)$$

Similarly, for refraction at the second surface and assuming A and B to be equidistant from the principal axis of the lens, we have

$$\frac{1/\mu}{v_1} - \frac{1}{v'} = \frac{1/\mu - 1}{r_2} + \frac{1/\mu - 1}{(1/\mu)^2}\left(\frac{1}{r_2} - \frac{1}{v'}\right)^2 \left(\frac{1}{r_2} - \frac{1/\mu + 1}{v'}\right)\frac{h^2}{2}$$

or $$\frac{1}{v_1} - \frac{\mu}{v'} = \frac{1 - \mu}{r_2} + (1 - \mu)\left(\frac{1}{r_2} - \frac{1}{v'}\right)^2 \left(\frac{\mu^2}{r^2} - \frac{\mu(\mu + 1)}{v'}\right)\frac{h^2}{2} \quad \ldots (ii)$$

Now in the last term in (*ii*) substitute for v' its approximate value given by

$$\frac{1}{v_1} - \frac{\mu}{v'} = \frac{1 - \mu}{r_2},$$

or $$\frac{1}{v'} = \frac{1}{\mu r_1} + \frac{\mu - 1}{\mu v_1}.$$

then $$\frac{1}{r_2}-\frac{1}{v'}=\frac{1}{\mu}\left(\frac{1}{r_2}-\frac{1}{v_1}\right)$$

and $$\frac{\mu^2}{r^2}-\frac{\mu(\mu+1)}{v'}=\frac{1}{r_2}-\frac{\mu+1}{v_1}.$$

Hence (*ii*) becomes

$$\frac{1}{v_1}-\frac{\mu}{v'}=\frac{1-\mu}{r_2}+\frac{1-\mu}{\mu^2}\left(\frac{1}{r_2}-\frac{1}{v_1}\right)^2\left(\frac{1}{r_2}-\frac{\mu+1}{v_1}\right)\frac{h^2}{2} \quad \text{... } (iii)$$

Adding (*i*) and (*iii*), we get

$$\frac{1}{v_1}-\frac{1}{u}=(\mu-1)\left(\frac{1}{r_1}-\frac{1}{r_2}\right)+\frac{h^2}{2}\cdot\frac{\mu-1}{\mu^2}$$

$$\left[\left(\frac{1}{r_1}-\frac{1+\mu}{u}\right)\left(\frac{1}{r_1}-\frac{1}{u}\right)^2-\left(\frac{1}{r_2}+\frac{\mu+1}{v_1}\right)\left(\frac{1}{r_2}-\frac{1}{v_1}\right)^2\right] \quad \text{... } (iv)$$

For paraxial rays, $$\frac{1}{v}-\frac{1}{u}=(\mu-1)\left(\frac{1}{r_1}-\frac{1}{r_2}\right) \quad \text{... } (v)$$

Hence substracting (*v*) from (*iv*), we have

$$\frac{1}{v_1}-\frac{1}{v}=\frac{h^2}{2}\cdot\frac{(\mu-1)K}{\mu^2}$$

where K stands for the expression within the chief brackets,

or $$v_1-v=-v^2\frac{h^2K}{2}\cdot\frac{\mu-1}{\mu^2}, \quad \text{... } [\because v_1\to v]$$

when $u\to\infty$, $v_1=f'$ the focal length for marginal rays, so that equation (*iv*) gives

$$\frac{1}{f'} = (\mu - 1)\left(\frac{1}{r_1} - \frac{1}{r_2}\right) + \frac{H^2}{2} \cdot \frac{\mu - 1}{\mu^2}$$

$$\left[\frac{1}{r_1^3} - \left(\frac{1}{r_2} - \frac{\mu+1}{f}\right)\left(\frac{1}{r_2} - \frac{1}{f}\right)^2\right]$$

where H is the semi-diameter of the lens, and f is substituted for f' in terms containing H^2.

$$\therefore \quad \frac{1}{f'} - \frac{1}{f} = \frac{H^2}{2} \cdot \frac{\mu - 1}{\mu^2}\left[\frac{1}{r_1^3} - \left(\frac{1}{r_2} - \frac{\mu+1}{f}\right)\left(\frac{1}{r_2} - \frac{1}{f}\right)\right]$$

or $$f' - f = -\frac{H^2}{2} f^2 \frac{\mu - 1}{\mu^2}\left[\frac{1}{r_1^3} - \left(\frac{1}{r_2} - \frac{\mu+1}{f}\right)\left(\frac{1}{r_2} - \frac{1}{f}\right)^2\right] \quad \ldots (2)$$

[$\because f'f = f^2$ approx.]

This gives the *longitudinal spherical aberration* of the lens for parallel rays.

To calculate the minimum spherical aberration for a direct pencil of parallel rays we differentiate the expression

$$\frac{1}{r_1^3} - \left(\frac{1}{r_1} - \frac{\mu+1}{f}\right)\left(\frac{1}{r_2} - \frac{1}{f}\right)^2,$$

whithin the chief brackets in equation (2) with respect to $1/r_1$ the curvature of the first surface subject to the condition that

$$\frac{1}{f} = (\mu - 1)\left(\frac{1}{r_1} - \frac{1}{r_2}\right)$$

and equate the resulting expression to zero. This gives

$$\frac{r_2}{f} = \frac{2(\mu - 1)(\mu + 2)}{2\mu^2 - \mu - 4}. \quad \ldots (2a)$$

Since $\frac{r_2}{f} = (\mu - 1)\left(\frac{r_2}{r_1} - 1\right)$, from the usual lens formula, we have by combining it with equation (2*a*)

$$\frac{r_2}{r_1} = 1 + \frac{r_2}{(\mu - 1)f} = \frac{\mu(2\mu + 1)}{2\mu^2 - \mu - 4} \qquad \text{... (2b)}$$

Whence further,

$$\frac{1}{r_1} = \frac{2\mu^2 + \mu}{2(\mu - 1)(\mu + 2)}\frac{1}{f}$$

and

$$\frac{1}{r_2} = \frac{2\mu^2 - \mu - 4}{2(\mu - 1)(\mu + 2)}\frac{1}{f}.$$

Putting these values in equation (2) we have minimum spherical aberration for parallel rays.

$$f' - f = -\frac{H^2}{2}\cdot\frac{1}{f}\mu(\mu - 1)$$

$$\left[\frac{(2\mu + 1)^3 - 9(-2\mu^2 - 2\mu + 1)}{8(\mu - 1)^2(\mu + 2)^3}\right]$$

$$= -\frac{H^2}{f}\cdot\mu\frac{4\mu - 1}{8(\mu - 1)^2(\mu + 2)}. \qquad \text{... (3)}$$

If $\mu = 1.5$, the minimum spherical aberration for an object at infinity is $-\frac{15}{14}\frac{H^2}{f}$, and by equation (2*b*) the ratio between the radii of curvature of the two surfaces of the lens $r_2/r_1 = -6$. For a plane spherical lens of the same focal length and placed with its curved surface opposite the object, the corresponding aberration is

$$-\frac{7}{6}\frac{H^2}{f}, \qquad \text{... (4)}$$

which is only about 8% higher.

Ray Tracing

The above mathematical treatment regarding spherical aberration, however, is somewhat tedious and, moreover, it can be easily said how the spherical aberration depends upon the various quantities mentioned. The *alternative* method is that of ray tracing in which the path of a ray through a given zone of the lens is faced by the application of Snell's law to each surface of the lens and the point where it crosses the axis, is determined.

The process no doubt involves a good deal of calculations for zones of different radii, lenses of different focal lengths, different radii of curvature of each face for a given focal length and for different refractive indices, but is more illustrative. [For sample calculation, the reader is referred to *Fundamentals of Optics* by Jenkins and White and *Applied Optics* and *Optical Design* by Conrady]. In this case graphs are drawn to find how the axial spherical aberration depends upon the above variables and how the conclusions arrived at assist us in the practical designing of lenses with minimum spherical aberration. One such graph showing the relation between the radius of the zone of the lens and axial spherical aberration. It shows that *axial spherical' aberration is proportional to the square of the radius of the zone.*

The next important point is to study how the curvature of the face of a lens presented to incident light for lenses of the same paraxial focal length but of different radii of curvature affects the spherical aberration. This alteration in the shape is known as bending the lens and the *shape factor of lens* is given by

$$x = \frac{r_2 + r_1}{r_2 - r_1} \qquad \ldots (5)$$

where r_1, r_2 are the radii of curvature of its two faces. For example, if the two radii of a converging meniscus are $r_1 = -16$ and $r_2 = -6$, it has a shape factor

$$x = \frac{-6-16}{-6+16} = -\frac{22}{10} = -2.2.$$

Shows the axial spherical aberration for lenses of different shapes but with the same power. Here, $f = 10$ cm, $d = 2$ cm, $\mu = 1.517$ and ray is incident from the left at a height 1 cm above the axis.

It will be seen that the aberration is large for the lenses in which the incident face is concave or slightly convex, *viz.*, the shape factor is negative. For shape factors between + 0.4 and + 1, the spherical aberration only varies slightly since it is close to a minimum which occurs at shape factor + 0.7, when the face at which the light leaves is slightly convex or plane. It then increases again as this face becomes more and more concave. We, therefore, see that by choosing the proper radii for the two surfaces of a lens, the spherical aberration can be reduced to a minimum but at no points it falls to zero.

By an extension of the condition of minimum deviation in a prism to a lens, we find that *spherical aberration increases approximately as the square of the deviation* and hence it is minimum when the total deviation of the ray is divided about equally between the two refractions. In any other case the reduced aberration at the surface where the deviation is least is more than counter-balanced by the increased aberration at the second surface.

For parallel light incident on a crown glass lens the condition for minimum spherical aberration is satisfied at a shape factor of about +0.7, not greatly different from the plano-convex lens for which $x = +10$, and hence the frequent use of plano-convex lenses in the construction of eye pieces and condensers etc.

Methods of Reducing Spherical Aberration : The following methods have been used for reducing spherical aberration :

(1) *By Means of Stops* : Obviously spherical aberration can be reduced by decreasing the aperture to a small region round the centre by putting a stop but this will reduce the definition and brightness of the image too, hence *alternatively*, it is preferable to intercept the central portion and allow only the peripheral rays to form the image. This improves the definition of the image (see Diffraction).

(2) *By the Use of Plano-convex Lenses* : For a lens of wide aperture, the spherical aberration may be reduced to a minimum by a suitable choice of the curvature of the faces of the lens, although it cannot be made to vanish completely. By the method of ray tracing discussed in the previous section we find that spherical aberration increases more and more quickly as the deviation becomes large. It is almost proportional to the square of the deviation and is minimum when the total deviation of the rays is divided equally between the two surfaces of the lens, that is, the angles of incidence and emergence must be equal. This condition is approximately satisfied when a plano-convex lens is used with the convex side facing the incident or emergent rays whichever are more parallel to the axis.

As an illustration let us consider a data from the '*Principles of Optics*' by Hardy and Perrin.

When the shape of the lens (μ =1.52) is double convex with each face having radius of curvature 10 cm for a zone of 2 cm diameter, the aberration is 15 mm. For a plano-convex lens of 10 cm focal length and diameter of the zone concerned equal to 2 cm, when the plane face is towards the distant object and the radius of curvature of the other face is 5.2 cm, the aberration is 4.4 mm; when the convex side is next to the same object, the aberration is 1.1 mm. So that, we find that the spherical aberration of a plano-convex lens when the convex side faces the incident parallel rays, is one-fourth of that when the plane surface faces it.

When, however, a plano-convex lens is used as the first lens in a microscopic objective, the rays diverge from a point very close to the surface of the lens and emerge almost as a parallel beam. In such a case the plane surface must be nearer the object.

(3) *By the Use of Crossed Lenses* : Mathematically it has been shown on page 113 equation (2*b*) that the form of a lens of a given focal length which will bring parallel rays of light to a focus with minimum of spherical aberration must be such that

$$\frac{r_2}{r_1} = \frac{2\mu^2 + \mu}{2\mu^2 - \mu - 4}.$$

where μ is the refractive index and r_1, r_2 are the radii of curvature of the first and the second surface of the lens respectively, so that if $\mu = 1.5$.

$$\frac{r_2}{r_1} = -6 \qquad \text{... (6)}$$

Lenses having this relationship between the curvatures of their surfaces in order to minimize spherical aberration are called crossed lenses.

Equation (6) leads to the conclusion that for minimum spherical aberration, the lens should be nearly plano-convex with the more convex side directed towards the distant source.

If $\mu = 1.686$, the denominator $(2\mu^2 - \mu - 4)$ becomes zero and hence

$$\frac{r_2}{r_1} = \infty$$

i.e., the lens should have its one surface plane.

It has also been shown in Equ. (4) that for an object at infinity the minimum spherical aberration of a lens with μ = 1.5 and $r_2/r_1 = -6$, is only 8% smaller than that of a planoconvex lens of the same material and hence the frequent use of plano-convex lenses in the manufacture of optical instruments.

(4) *By the Use of two Plano-convex Lenses Separated by a Finite Distance :* Here each lens produces the same deviation in a ray originally parallel to the axis and further since the deviation is spread over four refracting surfaces, each surface of the lens produces one-fourth of the total deviation and hence the spherical aberration is further reduced. Such a system of lenses is used in Huygens' eye-piece, the distance apart of the lenses being the difference in their focal lengths.

Let a be the distance apart of the two leases O_1 and O_2, with focal lengths f_1 and f_2 respectively. A ray PA parallel to the axis after refraction through O_1 is directed towards its second principal focus F_2. The deviation produced is equal to $\angle BAK = \angle BF_2 O_2 = \delta$, say.

Let the ray AB after refraction through O_2 meet the axis in F_1. If the deviation at B is the same as at A, then

$$\angle F_2BF_2 = \angle BF_2F_1 = \delta,$$

so that $$F_1F_2 = F_1B = F_1O_2,$$

or $$O_2F_1 = \frac{1}{2} O_2 F_2$$

Remembering that F_2 is the virtual object of the real image and applying the relation

$$\frac{1}{v} - \frac{1}{u} = \frac{1}{f_2} \qquad \text{for lens } O_2$$

we have $u = O_2F_2 = f_1 - a$; $v = O_2F_2 = \dfrac{1}{2} O_2F_2 = \dfrac{f_1 - a}{2}$.

$$\therefore \qquad \frac{2}{f_1 - a} - \frac{1}{f_1 - a} = \frac{1}{f_2}$$

or $$\frac{2-1}{f_1 - a} = \frac{1}{f_2},$$

or $$f_2 = f_1 - a$$

or $$a = f_1 - f_2. \qquad \text{... (7)}$$

This is the condition for minimum spherical aberration.

(5) *By Combining Suitable Concave and Convex Lenses* : For a concave lens the axial spherical aberration is negative since the marginal focus is on the opposite side of the paraxial focus to that from which the incident light is coming. So that by combining a concave lens with a convex one, the spherical aberration is partially compensated.

Coma

It is the name given to the comet-like appearance of the image of a point object situated just off the axis of a lens. This monochromatic aberration occurs even if the lens has been corrected for central spherical aberration and produces a sharp image of a particular point object on the axis. It is due to the fact that different zones of the lens being asymmetrical to the point just off the axis, do not give rise to a point image but ring images of increasing diameter and not concentrically situated and also having different transverse magnifications.

To understand coma let us consider the formation of the image of a small object $A A'$, placed normally to the axis of a lens working at full aperture. The central zone of the lens limited by the rays 1, 1' forms the image of A' at B'. The rays 2, 2' which make equal angles with $A' OB'$ come together at B_1' a little in front of the conjugate focal plane $B'B$ because the lens has a shorter focal length for non-paraxial rays. The point B_1' also does not lie on $A'OB'$ as the ray 2' is incident on the lens at a greater angle than ray 2 and, therefore, suffers greater deviation and hence is more inclined to $A'OB'$ than ray 2. The cone of rays starting from A' and inclined as rays 2, 2' will meet at B_1 approximately and if a screen is placed at BB', this cone will produce a bright circle on the screen after issuing out from B_1. The diameter of this circle will depend upon the distance of B_2 from $B'B$.

Similarly, the cone of rays inclined as rays 3, 3' will come to focus at B_2', further left of B_1' and more towards the axis and will produce a ring image on the screen of a greater diameter as B_2 is further removed from the screen. Considering the effect of the various zones of the lens in the image formation we find that the overall image of the point object consists of a series of circles, each representing the image from a different zone of the lens.

These circular patches will overlap and will produce an image in which the intensity is greater at B' and decreases as we go upwards towards B. The image gives the appearance of a comet with its tail and hence the name coma is assigned to this defect.

We also and that magnification is different for different zones of the lens. If the magnification for the outer rays is less than for the central rays, as in the case discussed above, the coma is said to be *negative*, while if the reverse is true, the coma is said to be *positive*.

Evidently, this defect will spoil the definition of the image on account of the asymmetrical form it gives to it and is serious in the case of optical instruments where large apertures are employed, such as photographic lenses, telescopes and more particularly in the case of microscopic objectives owing to the failure to fix the cross-wires in the correct position.

Abbe, a German optician has shown that coma can be eliminated if the sine condition is satisfied. According to this condition

$$\mu_1 h_1 \sin \alpha_1 = \mu_2 h_2 \sin \alpha_2,$$

where letters have their usual significance (of. equation 18).

We have seen above that coma is due to the variation in the lateral magnification of the image and to prevent it, therefore, this lateral magnification must be the same for all zones of the lens.

That is, $m = \dfrac{h_2}{h_1}$ should be constant.

From the sine condition, we have

$$\frac{\sin \alpha_1}{\sin \alpha_2} - \frac{h_2}{h_1} \times \frac{\mu_2}{\mu_1} = \text{constant.}$$

So to eliminate coma, Abbe's sine condition reduces to

$$\frac{\sin \alpha_1}{\sin \alpha_2} = \text{constant} \qquad \text{... (8)}$$

that is, the ratio of the sine of the slope of the object rays for a point on the axis and the sine of the corresponding slope for the image rays is constant. Such a surface is called an aplanatic surface.

Aplanatic Surface : This surface has the property of bringing all rays which are incident on it from a point source on the axis to a point focus on sent a spherical surface with axis AQ and centre of curvature C.

The surface separates air on the right from a medium of refractive index μ on the left. Consider an object placed at P at a distance from C equal to AC/μ where AC (=B) is the radius. Aplanatic surface. curvature of the surface.

A virtual image is formed at Q and it can easily be proved that

$$CQ = \mu AC.$$

Now, $$\frac{CP}{CM} = \frac{\sin i}{\sin \angle MPC} \quad \ldots (i) \text{ by sine formula}$$

But $$CP = \frac{CA}{\mu} = \frac{CM}{\mu}, \qquad [\because CA = CM]$$

or $$\frac{CP}{CM} = \frac{1}{\mu} = \frac{\sin i}{\sin r}$$

Hence, from (*i*), $\angle MPC = r$.

But $$\angle MPC = \angle MQP + \angle QMP,$$

or $$r = \angle MQP + r - i,$$

Hence in ΔCMQ, we have

$$\frac{CQ}{CM} = \frac{\sin \angle CMQ}{\sin \angle CQM} = \frac{\sin i}{\sin r} = \mu.$$

Consequently, $CQ = \mu CM = \mu R =$ a constant,

no matter at what angle the light is incident on the surface. In other words, P and Q are such points that rays diverging from P will, after refraction, appear to diverge from Q.

It will be noted that this surface satisfies for these points the sine condition referred to in *'coma'* above. These points are known as aplanatic foci and the surface an aplanatic surface. Such a surface eliminates both the spherical aberration and coma and is usefully employed in microscope objectives, where it is essential that as wide a pencil of light as possible from the object should enter the objective to have sufficient brightness in the image. An aplanatic meniscus lens, designed to satisfy these conditions. Here the object P lies at the centre of curvature C_1 of the first surface so that the rays fall normally and consequently their paths are not changed. On emerging from the second surface with its centre at C_2 on the axis, the rays bend away from the normal, i.e., the pencil of rays becomes less divergent and appears to proceed from the point Q, so that P and Q are aplanatic foci for the second surface.

For this to occur,

$$CP = \frac{O_2C}{\mu},$$

and
$$CQ = \mu O_2C.$$

If u be the distance of the object point from the first surface and v the distance of its image from the second surface, t the thickness O_1O_2 of the lens, then

$$u = O_2P - t = O_2C + CP - t$$

$$= O_2C + \frac{O_2C}{\mu} - t$$

$$= O_2C + \left(1 + \frac{1}{\mu}\right) - t = r_2\left(1 + \frac{1}{\mu}\right) - t$$

and
$$v = O_2Q = O_2C + CQ = O_2C + \mu O_2C$$

$$= O_2C\ (1 + \mu) = r_2(1 + \mu),$$

where $O_2C = r_2$ and μ is the refractive index of the material of the lens.

It is also possible to construct a divergent aplanatic meniscus lens. All rays proceeding towards C_1, the centre of curvature of the first surface will actually converge to Q independent of the point of incidence, if as in the case just discussed.

$$CP = O_2C/\mu$$

and $CQ = O_2C,$

where C is the centre of curvature of the second surface. Divergent aplanatic meniscus lens which can produce a real image of a real object cannot be constructed. For application of the principle of aplanatism see microscope objectives.

Astigmatism : This defect is due to those pencils of light which pass obliquely through the lens and occurs even if a stop is used to reduce spherical aberration and coma.

Consider a pair of rays proceeding from a point P well off the axis and above the plane of the paper. All rays from P in the tangential or a vertical plane PAB give rise to a point image at Q_1 on the central ray POQ_1 but the rays from P in the horizontal or sagittal plane PCD form a point image at Q_2 also on the central ray but much farther than Q_1.

The convergent beam making for Q_2 as it passes Q_1, gives rise to a line image T parallel to CD called the first focal line. Similarly, the beam making for Q_1 spreads out again giving rise to another line image S parallel to AB called the second focal line. It can be shown that all the rays between these two planes pass through these lines after emergence from the lens. The result is that nowhere a sharp image of P is obtained, and the thin pencil, which diverges from two short lines mutually at right angles, is said to be astigmatic. Hence, the name astigmatism for this type of aberration.

In between these two focal lines, called the first and second focal lines, the image is approximately disc-shaped and denotes the circle of least confusion which is the nearest approach to the image of the object. The distance between these two lines is called the astigmatic difference. This difference increases with the obliquity of the rays and is reduced to zero for a paraxial beam. The image of a plane object perpendicular to the axis is thus formed of two curved surfaces..

On the axis, where the two surfaces come together the astigmatic difference is nil; away from the axis it increases approximately as the square of image height. The astigmatism is said to be *positive* when the *T* surface lies to the left of *S* For a divergent lens the astigmatism is opposite. So it is obvious that by making use of convex and concave lenses of suitable focal lengths separated from each other, it is possible to get rid of astigmatic difference. Such a lens is called an anastigmat and is largely employed in the construction of photographic camera objectives in which wide angled pencils have to be used.

Curvature of the Field : When an extended plane object is viewed through a convex lens, say, we find that the image is not a flat one. Generally, the centre of the field will be in sharp focus but the edges will be quite blurred. This is because the focal length of the lens for a paraxial beam is greater than for oblique pencils. Thus, the image of points situated off the axis is drawn farther from the lens than that of the central paraxial parts in the case of a virtual image and is drawn close to the lens in the case of a real image.

This aberration arising from the oblique centric rays falling on the lens is called the curvature of the field. It occurs even if the aperture is cut down by a suitable stop so as to reduce central spherical aberration, coma, and astigmatism.

The defect can be easily demonstrated when a squared paper is seen through a convex lens of short focus. We will find that in the case of a virtual image, *i e.,* when convex lens is used as a magnifier, the lines would appear to curve away from each other near the edges of the lens as if they are stretched out and are only parallel over quite a small central part. When, however, a real image is seen, opposite effects occur, *viz.,* the lines curve towards each other at the edges as if the ends are compressed;.

In the case of a divergent lens, where (he image of a real object is always virtual, the parts of the image, remote from the axis, are drawn up towards the lens, and the image is, therefore, concave towards the lens.

It can be shown theoretically that for a system of thin co-axial lenses the curvature in given by

$$\frac{1}{R} = \sum \frac{1}{\mu_n f_n} \qquad \dots (9)$$

where μ_n is refractive index of the material of the nth lens and f_n its focal length, so that for no curvature R is ∞.

$$\therefore \qquad \sum \frac{1}{\mu_n f_n} = \frac{1}{\infty} = 0 \qquad \dots (10)$$

In the case of two lenses, placed in air, the condition reduces to

$$\frac{1}{\mu_1 f_1} + \frac{1}{\mu_2 f_2} = 0$$

$$\text{or} \qquad \mu_1 f_1 + \mu_2 f_2 = 0 \qquad \dots (11)$$

where f_1, f_2 are the focal lengths of the lenses and μ_1 and μ_2 are their respective refractive indices.

This condition for no curvature of the field was proved by Petzval and is, therefore, called Petzval's condition. It holds whether the lenses are in contact or separated. Since μ_1, μ_2 are positive, the focal lengths of the constituent lenses f_1, f_2 must be of opposite sign. Thus, by combining suitable convergent and divergent lenses the image may be made flat. This condition should be satisfied by photographic objectives because of the wide field they must cover.

Meniscus Lens : Curvature of the field may also be corrected for a single lens by using a stop in the proper position on the axis of the lens. It is located in front of a meniscus lens so that the light is incident on the concave surface. By arranging the aperture stop at about $f/11$, the rays from each point object can be made to pass through the lens in such a way that the *primary* (tangential) and the *secondary* (sagittal) image surfaces T and S have equal and opposite curvatures, thus resulting in a flat image. Sharp focus is, however, obtained in the centre, astigmatism producing a blurring at the margins. It is generally used in cheapest box cameras and being uncorrected for chromatic aberration it does not give very good photographs.

Distortion : There is another defect arising from the variation in the lateral magnification of different parts due to the differences in the distances of the object points from the axis and thus producing lack of geometrical similarity between the object and its image. It is called distortion. The magnitude of this defect is the greater the distance of the object points from the axis. There are two common forms of distortion that we meet within the case of lens images. Thus, if the real image of a rectangular piece of wire gauze is formed on a screen with a convex lens, the appearance is barrel shaped owing to the magnification of the marginal parts being less than the central ones. This is known as barrel-shaped distortion.

If, however, a virtual image of the same piece of wire gauze is seen by the naked eye, it appears of the form and is called pin-cushion distortion corresponding to a greater magnification at the borders. But a convex lens generally exhibits both forms of distortion in the formation of real images depending upon the position of the stops which arc generally employed in the construction of several optical instruments. If a stop is placed between the object and the lens, rays from object points like Q will pass through the central portion of the lens and give rise to the image at Q' while the rays from off axis object points like P will pass through the lower half of the lens. Hence, for the extremities of the object, the object distance *u* will be greater than the paraxial values and the ratio of the image to the object distance, *viz*, the *magnification*

$$\left(m = \frac{v}{u} = \frac{f}{u+f}\right)$$

will be less than for object points nearer the axis. Therefore, the image will appear to be pulled in at the corners, *i.e.*, if the image for the paraxial rays is at P' that due to the extreme rays will be at P'_1. Thus, if the object is a square and perpendicular to the axis of the lens, the image will be barrel shaped as shown in continuous lines in the right-hand side. If the same stop is placed on the image side of the lens, the rays forming the image of the extremities of the object have a greater value than tor the paraxial rays, hence the magnification

$$\left\{m = \frac{(f-v)}{f}\right\}$$

is increased for the outer parts of the field. The image of a square will thus appear to be palled out at the corners and the result is pincushion distortion, as shown on the right-hand side, in which the point P' refers to the image by paraxial rays and P_1' due to extreme rays.

This defect is due to the principal ray and it is found to persist even if the width of the stop is infinitely small and the spherical aberration has been corrected. So the defect cannot be ascribed as being due to a difference in the focal lengths for the marginal and the paraxial rays.

Rapid Rectilinear : What we have learnt about disortion suggests that it can be corrected by combining two similar lenses placed a certain distance apart with a stop midway between them. In this way a system is obtained in which pin cushion distortion due to the first is compensated by the barrel-shaped distortion due to the second so that the resultant image is flat. Such a corrected lens system is called an orthoscopic doublet or rapid rectilinear lens.

The combination can be made free from chromatism by using an achromatic doublet for each component. Such systems of lenses can be conveniently used as photographic objectives upto an aperture of about $f/8$, thus admitting twice as much light as in the case of a single meniscus-type.

Chromatic Aberration

Dispersion Produced by a Prism : We know that beam of white light is refracted through a prism it is split up into its constituent colours red, orange, yellow, green, blue, indigo and violet, each having a different refrangibility. This separation of colours by a prism is known as dispersion and the band of colours is called the spectrum. The angle between the directions of any two colours measures the angular dispersions (angular separation) between them and is equal to the difference between the deviation for the two colours.

If d_b and d_r be the angles of deviations for the extreme blue and red rays respectively and (and the corresponding refractive indices for a prism with a small refracting angle A, then

$$d_b = (\mu_b - 1)A$$

and $$d_r = (\mu_r - 1) A$$

$\therefore$ angular separation between these two colours

$$d_b - d_r = (\mu_b - 1) A - (\mu_r - 1) A$$

$$= (\mu_b - \mu_r) A.$$

If d be the angle of deviation for the mean ray, say the yellow light and its refractive index, then

$$\delta = (\mu - 1) A.$$

The ratio between the angular separation and the mean deviation when light is dispersed at minimum deviation by a prism of small angle is called the 'dispersive power of the medium of which the prism is made. Denoting it by ω, we have

$$w = \frac{d_b - d_r}{d} = \frac{(\mu_b - \mu_r)A}{(\mu - 1)A}$$

$$= \frac{\mu_b - \mu_r}{\mu - 1} = \frac{\delta\mu}{\mu - 1} \quad \text{... (1)}$$

in the language of differential calculus. It is independent of the angle of the prism and is constant for its material.

Angular and Chromatic Dispersions of the Material of a Prism : If the angular separation between two neighbouring colours of wavelengths λ and λ + δλ is denoted by δθ, then the ratio δθ/δλ (= δθ/δλ) measures the angular dispersion of the material of the prism. It depends upon the angle of deviation and hence the position of the prism. Let us calculate its value in the position of *minimum* deviation. The refractive index of the material of a prism is given by

$$\mu = \frac{\sin(A + \theta)/2}{\sin A/2} \quad \text{... (i)}$$

where A is the angle of the prism and θ the angle of minimum deviation.

Differentiating it with respect to θ, we have

$$d\mu = \frac{\frac{1}{2}\cos(A+\theta)/2}{\sin A/2} d\theta$$

or $$\frac{d\mu}{d\lambda} = \frac{\cos(A + \theta)/2}{2\sin A/2} \frac{d\theta}{d\lambda}, \quad \text{... (2)}$$

$$\text{whence} \qquad \frac{d\theta}{d\lambda} = \frac{2 \sin A/2}{\cos (A+\theta)/2} \frac{d\mu}{d\lambda}$$

$$= \frac{2 \sin A/2}{\sqrt{1-\mu^2 \sin^2 A/2}} \frac{d\mu}{d\lambda}, \qquad \ldots (3)$$

since by (*i*) $\cos (A + \theta)/2 = \sqrt{1-\mu^2 \sin^2 A/2}$:

The ratio $d\mu/d\lambda$ is called the chromatic dispersion of the material of the prism. Hence, the angular dispersion of the material of a prism depends upon (*a*) the angle of the prism and (*b*) the material of the prism. To calculate its value for a given region of the spectrum, the prism is set in the minimum deviation position on the spectrometer table and μ is determined for a set of different wavelengths and a curve obtained between μ and δ. From the curve we can find the value of $d\mu/d\lambda$ by drawing a tangent at a particular point and hence that of $d\theta/d\lambda$ can be calculated for a given region of the spectrum with the help of equation (3). It may be noted that for radiations of wavelengths far removed from those of absorption bands of the medium, the refractive index increases regularly with decreasing wavelength or increasing frequencies, the dispersion is then said to be normal.

Irrational Dispersion : Suppose we have two prisms of different types of glass with such refracting angles that the angular separation between the extreme colours, say, lines *A* and *H* is exactly the same for each. It is found that angular separation between *A* and some intermediate line such as *D* in the spectrum produced by each prism is not the same as may be expected otherwise, *i.e.*, the intermediate lines do not generally coincide. Besides this, we notice that in the spectrum produced by a prism violet is much more spread out than the red. Such features of prismatic spectra are termed irrational dispersion.

For example, if μ, $\mu + \delta\mu$ are the refractive indices for two rays A and B and n, $n + \delta n$ are the refractive indices for two rays C and D in crown glass and μ', $\mu' + \delta\mu'$ and n', $n' + \delta n'$ denote the same quantities for flint glass, the ratio of the dispersive powers for crown glass is

$$\frac{\delta\mu}{\mu-1} : \frac{\delta n}{n-1}$$

and for flint glass it is

$$\frac{\delta\mu'}{\mu'-1} : \frac{\delta n'}{n'-1}$$

and these two ratios are not equal.

This is due to the molecular structure of the prism being such as to impede waves of shorter wavelength more than those of longer wavelength, and secondarily to the vibration of the molecular systems reacting on the ether undulations and tending to alter their periods.

Achromatic Prisms : When white light passes through a prism, deviation and dispersion both are produced. Since the dispersive powers of prisms of different materials are different, it is possible to design a combination of prisms so as to eliminate dispersion while a resultant deviation remains. Such a combination is said to be achromatic.

Suppose we have two prisms of crown and flint glasses with refracting angles equal to A and A' respectively and with their refracting edges pointing in opposite directions. If their refractive indices for red, blue and yellow light be μ_r, μ_b, μ and μ_r', μ_b', μ' respectively we have with usual notation

$$d_b - d_r = (\mu_b - \mu_r)A \qquad \text{for crown glass}$$

and

$$d_b' - d_r' = (\mu_b' - \mu_r')A' \qquad \text{for flint glass.}$$

If the system is to be achromatic $(d_b - d_r) - (d_b' - d_r') = 0$,

or $\quad (\mu_b - \mu_r)A - (\mu_b' - \mu_r')A' = 0$

or $$(\mu_b - \mu_r)A = (\mu_b' - \mu_r')A',$$

or $$\frac{A}{A'} = \frac{\mu_b' - \mu_r'}{\mu_b - \mu_r} = \frac{\delta\mu'}{\delta\mu} \quad \text{... (4)}$$

Since the dispersion due to the flint glass is greater than that due to crown glass, for achromatism the angle of the flint glass prism will be smaller than that of the crown glass prism.

If d and d' be the deviations of the mean rays after passing through the crown and flint glass prisms, then

$$d = (\mu - 1)A,$$

and $$d' = (\mu' - 1)A'.$$

The resultant deviation of the mean ray (yellow light) is given by

$$\phi = d - d' = (\mu - 1)A - (\mu' - 1)A'$$

$$= \frac{(\delta\mu)A}{\omega} - \frac{(\delta\mu')A'}{\omega'} \left[\omega = \frac{\delta\mu}{\mu - 1}, \quad \omega' = \frac{\delta\mu'}{\mu' - 1}\right]$$

where ω and ω' are the dispersive powers concerned. Since the system is achromatic $(\delta\mu)A = (\delta\mu')A'$ by equation (4) above, so that

$$\phi = \frac{(\delta\mu)A}{\omega} - \frac{(\delta\mu')A'}{\omega'} = (\delta\mu)A\left(\frac{1}{\omega} - \frac{1}{\omega'}\right) \quad \text{... (5)}$$

$$\therefore \quad A = \frac{\phi}{(\delta\mu)\left(\frac{1}{\omega} - \frac{1}{\omega'}\right)} \quad \text{... (6)}$$

with a similar expression for A'. Thus, it is possible to calculate the angles of component prisms so that the system shall be achromatic and the deviation of the mean ray has a given value ϕ.

$$\text{Since} \qquad \omega' > \omega; \left(\frac{1}{\omega} - \frac{1}{\omega'}\right) \text{ is + ve.}$$

Hence, the achromatic prism produces mean deviation in the same direction as produced by the crown glass prism.

Now, the dispersive powers of some kinds of flint and crown glasses are 0.053 and 0032 respectively. The crown glass prism with $\mu = 1.52$ and refracting angle 60° will in the position of minimum deviation produce a deviation of 38° 55' and an angular dispersion of about 0.032 × 38° 55' or 1° 15'. A flint glass prism with $\mu = 1.58$ and refracting angle 37° 26' will have the same dispersive effect in the minimum deviation position with the angle of deviation equal to 23° 30'. It is obvious, therefore, that when such prisms are placed in the position of minimum deviation with their refracting angles in opposite directions, the dispersion will be neutralized, whereas total deviation of 38° 55' — 23° 30' or 15° 25' will remain outstanding. The emergent light will not, however, be totally free from colour as the spectrum formed by a flint glass prism is relatively more extended in the blue part than the crown glass spectrum. So in practice two colours in the brightest part of the spectrum are united and the rest are allowed to take their chance.

If the two prisms are made of the same material, then $\mu_b = \mu_b'$ and $\mu_r = \mu_r'$ and hence A must be equal to A', *i.e.*, $d - d' = 0$, or there is no deviation of the emergent beam, which will, therefore, be parallel to the incident beam, and the combination would form a parallel-sided plate of the same material and both the dispersion and the deviation will he annulled.

A more useful combination of the prisms in the one in which no deviation is produced for the mean rays of the spectrum while the dispersion is left outstanding. This arrangement is employed in the construction of direct vision spectroscope.

Direct Vision Spectroscope : The direct vision spectroscope is a simple instrument designed merely for the qualitative study of the colours and the positions of the spectral lines due to a source in its line of sight. This spectroscope, as its name implies, gives a fair amount of dispersion but no deviation of the *mean* or the *standard* ray. It consists of alternate crown and flint glass prisms cemented together and arranged with their refracting angles in opposite directions so that flint glass prisms produce equal and opposite deviation to that produced by crown glass prisms, care being taken that the system begins and ends with a crown glass prism. Since the dispersive power of flint is greater than that of crown glass it follows that dispersion will remain outstanding. If the prisms are of small angles, then for the mean ray

$$d = (\mu - 1)A \text{ for one set of prisms,}$$

or $$d = (\mu' - 1)\ A' \text{ for the second set,}$$

A being the sum of the refracting angles of one set and A' that of the other.

For no resultant deviation for the middle part of the spectrum

$$d - d' = 0.$$

$$\therefore \quad (\mu - 1)\ A - (\mu' - 1)\ A' = 0$$

or $$\frac{A}{A'} = \frac{\mu' - 1}{\mu - 1}. \qquad \text{... (7)}$$

The dispersion, however, would be outstanding as for one set of prisms

$$d_b - d_r = (\mu_b - 1)\ A - (\mu_r - 1)\ A = (\mu_b - \mu_r)A$$

and for the second set of prisms

$$d_b' - d_r' = (\mu_b' - 1)\,A' - (\mu_r' - 1)\,A' = (\mu_b' - \mu_r')A'$$

$\therefore$ resultant dispersion

$$= (d_b - d_r) - (d_b' - d_r')$$

$$= (\mu_b - \mu_r)A - (\mu_b' - \mu_r')A'$$

$$= \frac{\mu_b - \mu_r}{\mu - 1}(\mu - 1)A - \frac{\mu_b' - \mu_r'}{\mu - 1}(\mu' - 1)A'$$

$$= \omega d - \omega' d' = (\omega - \omega')d \qquad [\because d = d'] \ldots(8)$$

where ω and ω' are the dispersive powers of the material of the two sets of the prisms.

Since $(\omega - \omega')$ is not zero, the dispersion remains outstanding.

Also since $\omega' > \omega$, $(\omega - \omega')$ is negative. Hence the order of colours in the resultant dispersion produced by the direct vision spectroscope is opposite to that obtained by dispersion due to the set of crown glass prisms.

A simple form of direct vision spectroscope in which three crown glass and two flint glass prisms of suitable angles are mounted alternately in a brass tube and held in position by cork pieces. They are arranged with their refracting angles in opposite directions so that the middle part of the spectrum suffers no deviation. The prisms are cemented together by Canada balsam to avoid loss of intensity due to reflection at the faces. There is an *adjustable* slit S placed parallel to the refracting edges of the component prisms and an *achromatic lens* at one end of the tube and an *eye-piece* at the other.

A narrow beam of white light passing through the slit is rendered parallel by the lens L_1. It then traverses the combination of the prisms, where it is dispersed. It may be noted that the rays cross one another in the second crown glass prism. They emerge out as a system of parallel beam which falling on another converging lens L_2 is brought to focus in its second focal plane to form a spectrum *VR*. This spectrum is next examined by the eye-piece *E* which acts as a simple magnifying glass By directing the spectroscope towards a source of light, say a star, a qualitative study of the spectra can be made, but as the resolving power is small, lines close to each other cannot be easily separated. It is well-suited to the study of flame spectra.

Chromatic Aberration of a Lens : If a parallel beam of white light is incident on a convex lens, on transmission through it we find that the red rays in the transmitted beam are brought to focus at a point farther than the blue rays. Thus a single lens produces coloured images of an object which are at different positions This prismatic action of a lens is known as chromatic aberration and is due to the fact that the refractive index of a substance is different for different colours increasing as we pass from red to blue. Thus, since

$$\frac{1}{f} = (\mu - 1)\left(\frac{1}{r_1} - \frac{1}{r_2}\right)$$

where the letters have their usual significance, it follows that f has different values for different colours. So if f_r, μ_r, denote the focal length and refractive index for red rays, and f_b, μ_b the corresponding quantities for blue rays, we have

$$\frac{1}{f_r} = (\mu_r - 1)\left(\frac{1}{r_1} - \frac{1}{r_2}\right)$$

and
$$\frac{1}{f_b} = (\mu_b - 1)\left(\frac{1}{r_1} - \frac{1}{r_2}\right)$$

Since μ_b is greater than μ_r, f_b is less than f_r, and the difference $f_r - f_b$ is a measure of the axial chromatic aberration of a lens for parallel rays.

Now, $$\frac{1}{f_r} = (\mu_r - 1)\left(\frac{1}{r_1} - \frac{1}{r_2}\right)$$

$$= \frac{\mu_r - 1}{\mu - 1}(\mu - 1)\left(\frac{1}{r_1} - \frac{1}{r_2}\right) = \frac{\mu_r - 1}{\mu - 1} \cdot \frac{1}{f}$$

and $$\frac{1}{f_b} = (\mu_b - 1)\left(\frac{1}{r_1} - \frac{1}{r_2}\right)$$

$$= \frac{\mu_b - 1}{\mu - 1}(\mu - 1)\left(\frac{1}{r_1} - \frac{1}{r_2}\right) = \frac{\mu_b - 1}{\mu - 1} \cdot \frac{1}{f}$$

$$\therefore \quad \frac{1}{f_b} - \frac{1}{f_r} = \frac{\mu_b - 1}{\mu - 1} \cdot \frac{1}{f} - \frac{\mu_r - 1}{\mu - 1} \cdot \frac{1}{f} = \frac{\mu_b - \mu_r}{\mu - 1} \cdot \frac{1}{f},$$

or $$\frac{f_r - f_b}{f_b f_r} = \frac{\mu_b - \mu_r}{\mu - 1} \cdot \frac{1}{f} \qquad \ldots [\because f_b f_r = f^2]$$

$$\therefore \quad f_r - f_b = \frac{\mu_b - \mu_r}{\mu - 1} f = \omega f, \qquad \ldots (9)$$

where ω is the dispersive power of the material of the lens.

Thus, the 'axial chromatic aberration' of a lens for parallel rays is equal to its mean focal length multiplied by the dispersive power of the material of the lens.

Besides axial chromatic aberration there is another chromatic defect resulting from the variation Of focal length of a lens with colour. This gives rise to a difference of magnification which shows the red and violet images of an off-axis object point. This defect is known as lateral chromatic aberration and is measured by the vertical difference in heights ($h_r - h_v'$) of the two images. Because these aberrations spoil the definition of the image, correction for both lateral and longitudinal chromatism is of considerable importance.

Achromatic Combination of two Thin Lenses in Contact : When a convex lens is used to obtain the real image of an object on the screen, the focus for blue rays is nearer the lens than that for the red rays. Similarly, if a convergent beam of white light falls on a concave lens so that a real image is produced, red image is nearer the lens than the blue. This suggests that if suitable convex and concave lenses are combined, the dispersion between two colours produced by one can be annulled by the other. To do so we combine a convex lens of crown glass which is more strongly *convergent* for blue than for red rays and a concave lens of flint glass which is more strongly *divergent* for blue than for red light, so that the crown glass lens of large positive power produces the same dispersion as the flint glass lens with a lower negative power.

The power of the combination is, therefore, positive while the dispersion is neutralised bringing all the colours to approximately the same focus. Such a combination is said to be an achromatic doublet. The lenses are cemented together with canada balsam to avoid loss of light by reflection. In order that this may be possible the radius of curvature of the second surface of the convex lens is made the same as that of the first surface of the second lens. It may be noted that on account of irrationality of dispersion cannot make a combination which is perfectly achromatic but Herchel has suggested that "the exact union of the kinds of light denoted by the D and F lines of the spectrum will ensure the approximate union of all the rest better on the whole than if we aimed at uniting the extremes of the spectrum and a far greater concentration of light will be produced". Let us now derive the condition for achromatism of two thin lenses in contact.

Let f and f' be the mean focal lengths of the crown and flint glass lenses, μ_b, μ_r and μ the refractive indices for blue, red and yellow light for the crown glass and μ'_b, μ'_r and μ' the corresponding values for the flint glass. If F_r, F_b be the focal lengths of the combination for the red and blue ray, we have

$$\frac{1}{F_r} = \frac{1}{f_r} + \frac{1}{f'_r}$$

$$= (\mu_r - 1)\left(\frac{1}{r_1} - \frac{1}{r_2}\right) + (\mu'_r - 1)\left(\frac{1}{r'_1} - \frac{1}{r'_2}\right)$$

$$= \frac{\mu_r - 1}{\mu - 1}(\mu - 1)\left(\frac{1}{r_1} - \frac{1}{r_2}\right)$$

$$+ \frac{\mu'_r - 1}{\mu' - 1}(\mu'_r - 1)\left(\frac{1}{r'_1} - \frac{1}{r'_2}\right)$$

$$= \frac{\mu_r - 1}{\mu - 1} \cdot \frac{1}{f} + \frac{\mu'_r - 1}{\mu' - 1} \cdot \frac{1}{f'} .$$

Similarly $$\frac{1}{F_b} = \frac{\mu_b - 1}{\mu - 1} \cdot \frac{1}{f} + \frac{\mu'_b - 1}{\mu' - 1} \cdot \frac{1}{f'}$$

If the combination is to be achromatic, $F_b = F_r$,

whence $$\frac{\mu_b - 1}{\mu - 1} \cdot \frac{1}{f} + \frac{\mu'_b - 1}{\mu' - 1} \cdot \frac{1}{f'} = \frac{\mu_r - 1}{\mu - 1} \cdot \frac{1}{f} + \frac{\mu'_r - 1}{\mu' - 1} \cdot \frac{1}{f'}$$

By transposing and simplifying, we have

$$\frac{\mu_b - 1}{\mu - 1} \cdot \frac{1}{f} + \frac{\mu_r - 1}{\mu - 1} \cdot \frac{1}{f} + \frac{\mu'_b - 1}{\mu' - 1} \cdot \frac{1}{f} - \frac{\mu'_r - 1}{\mu' - 1} \cdot \frac{1}{f'} = 0,$$

or $$\frac{\mu_b - \mu_r}{\mu - 1} \cdot \frac{1}{f} + \frac{\mu'_b - \mu'_r}{\mu' - 1} \cdot \frac{1}{f'} = 0$$

Denoting the dispersive powers of the two lenses by ω and ω' we have

$$\frac{\omega}{f}+\frac{\omega'}{f'} = 0$$

or
$$\frac{f}{f'} = \frac{\omega}{\omega'}. \qquad \text{... (10)}$$

That is (*i*) the ratio of the focal lengths of the two lenses is numerically equal to the ratio of the dispersive powers of their materials and (*ii*) since ω and ω′ are both positive, the focal lengths of the two lenses must be of opposite sign.

Alternative Method : The above condition may be more easily deduced by the method of differential calculus.

For a thin lens

$$\frac{1}{f} = (\mu-1)\left(\frac{1}{r_1}-\frac{1}{r_2}\right).$$

The variables are f and μ, so that by differentiating, we have

$$-\frac{df}{f^2} = d\mu\left(\frac{1}{r_1}-\frac{1}{r_2}\right)$$

$$= \frac{d\mu}{\mu-1}(\mu-1)\left(\frac{1}{r_1}-\frac{1}{r_2}\right)$$

or
$$-\frac{df}{f} = \frac{d\mu}{\mu-1} = \omega. \qquad \text{... (}i\text{)}$$

where ω is the dispersive power of the medium. Now let the focal lengths of the constituent lenses be f and f' and the dispersive powers of their materials ω and ω′. Then if the focal length of the combination is F, we have

$$\frac{1}{F} = \frac{1}{f}+\frac{1}{f'}$$

By differentiating, we get

$$-\frac{dF}{F^2} = -\frac{df}{f^2} - \frac{df'}{f'^2}$$

$$= -\frac{df}{f}\cdot\frac{1}{f} - \frac{df'}{f'}\cdot\frac{1}{f'}$$

$$= \frac{\omega}{f} + \frac{\omega'}{f'} \qquad \text{... by equation } (i)$$

For achromatism, $dF = 0$, that is, the focal length is the same whatever the colour of light,

$$\therefore \qquad \frac{\omega}{f} + \frac{\omega'}{f'} = 0$$

or

$$\frac{f}{f'} = \frac{\omega}{\omega'}. \qquad \text{... } (ii)$$

In order, therefore, to design an achromatic doublet of focal length F, the focal lengths of the constituent lenses must satisfy the above relation in addition to the relation,

$$\frac{1}{F} = \frac{1}{f} + \frac{1}{f'} \qquad \text{... } (iii)$$

Knowing ω, ω' and F, the magnitudes of f and f' can be found by solving equations (*ii*) and (*iii*).

Since the combination is to be equivalent to a convergent lens, f will be found to be less than f and since for achromatism $\frac{\omega}{f} = \frac{\omega'}{f'}$ we conclude that $\omega < \omega'$ that is, the material of the convex lens must have a lower dispersive power.

From the lens manufacturers' point of view, however, the radii of the surfaces of the two lenses have to be calculated and to secure efficient functioning the lenses must be in good optical contact. The curvature of the first surface of the flint glass lens must be the same as that of the second surface of the crown glass lens that is, $r'_1 = r_2$ If in addition the spherical aberration of the system is to be minimum, the free surface of the flint glass should be plane, that is $r_2' = \infty$, so that the focal lengths of the two lenses in contact in terms of the radii of curvature become

$$\frac{1}{f} = (\mu - 1)\left(\frac{1}{r_1} - \frac{1}{r_2}\right) \quad \text{... } (iv)$$

$$\frac{1}{f'} = (\mu' - 1)\left(\frac{1}{r_1'} - \frac{1}{r_2'}\right) = (\mu' - 1)\left(\frac{1}{r_2} - \frac{1}{\infty}\right) \quad \text{... } (v)$$

Since $r'_1 = r_2$ and $r'_2 = \infty$.

So knowing f' and μ', r_2 can be calculated from equation (*v*) Setting its value in equation (*iv*), r_1, may be calculated, so that the two lenses are completely designed.

Example : *Given the dispersive powers of the crown and flint glasses as 0.2 and 0.4 respectively, find the focal lengths of the components of an achromatic doublet of focal length 20 cm.*

Let f, f' be the focal lengths of the crown and flint glass lenses and F, the focal length of the combination. Then

$$\frac{1}{F} = \frac{1}{f} + \frac{1}{f'} \quad \text{... } (a)$$

and
$$\frac{f}{f'} = -\frac{\omega}{\omega'} \quad \text{... } (b)$$

From (*b*)
$$f = -f'\frac{\omega}{\omega'} = -f' \times \frac{0.2}{0.4} = -\frac{f'}{2},$$

From (*a*)
$$\frac{1}{20} = \frac{2}{f'} + \frac{1}{f'} = \frac{-2+1}{f'} = -\frac{1}{f'}.$$

$$\therefore \qquad f' = -20 \text{ cm}$$

and
$$f = -\frac{-20}{2} = +10 \text{ cm.}$$

Example : *The refractive indices for sodium light for crown and flint glasses are 1.510 and 1.650 respectively and faces in contact of the two components of an achromatic doublet have radii of curvature 15 cm. Calculate the radii of curvature of other faces of the lenses using the data of the previous example.*

If the combination is arranged so that convex lens is nearer the object, then

$$\frac{1}{f} = (\mu - 1)\left(\frac{1}{r_1} - \frac{1}{r_2}\right)$$

Here, $f = +10$ cm; $\mu = 1.510$; $r_2 = -15$ cm.

$$\therefore \qquad \frac{1}{10} = 0.51\left(\frac{1}{r_1} + \frac{1}{15}\right)$$

$$= \frac{51}{100\, r_1} + \frac{51}{1500} = \frac{51}{100\, r_1} + \frac{17}{500}.$$

$$\therefore \qquad \frac{51}{100\, r_1} = \frac{1}{10} - \frac{17}{500} = \frac{50 - 17}{500} = \frac{33}{500}$$

or
$$\frac{51}{r_1} = \frac{33}{5},$$

or
$$r_1 = \frac{51 \times 5}{33} = \frac{85}{11} = 7.73 \text{ cm.}$$

Similarly, for the concave lens

$$-\frac{1}{20} = 0.65\left(-\frac{1}{15} - \frac{1}{r_2'}\right) = -0.65\left(\frac{1}{15} - \frac{1}{r_2'}\right)$$

or
$$\frac{1}{15} - \frac{1}{r_2'} = \frac{1}{20 \times 0.65} = \frac{1}{13}.$$

or $$\frac{1}{r'_2} = \frac{1}{13} - \frac{1}{15} = \frac{15-13}{195} = \frac{2}{195}.$$

∴ $$r'_2 = \frac{195}{2} = 97.5 \text{ cm}.$$

Example : *An achromatic objective of focal length 30 cm is to comprise thin lenses cemented together with canada balsam and made from hard crown and dense flint glasses respectively. The free surface of the divergent lens is plane, calculate the radii of curvature of both the lenses.*

	crown	flint
μ_{red}	1.514	1.644
μ_{blue}	1.524	1.664

The four relevant equations are —

$$\frac{1}{F} = \frac{1}{f} + \frac{1}{f'}, \quad \ldots (i)$$

and $$\frac{f}{f'} = -\frac{\omega}{\omega'}, \quad \ldots (ii)$$

$$\frac{1}{f} = (\mu - 1)\left(\frac{1}{r_1} - \frac{1}{r_2}\right), \quad \ldots (iii)$$

and $$\frac{1}{f'} = (\mu' - 1)\left(\frac{1}{r'_1} - \frac{1}{r'_2}\right) \quad \ldots (iv)$$

Here, $F = 30$ cm.

For crown glass average refractive index (μ)

$$= \frac{1.514 - 1.524}{2} = \frac{3.038}{2} = 1.519.$$

$$\text{Dispersive power } (\omega) = \frac{1.524 - 1.514}{1.591 - 1} = \frac{0.01}{0.519} = \frac{10}{519}.$$

For flint glass average refractive index (μ')

$$= \frac{1.664 + 1.664}{2} = \frac{3.308}{2} = 1.654$$

$$\text{Dispersive power } (\omega') = \frac{1.664 + 1.664}{1.654 - 1} = \frac{0.02}{0.654} = \frac{10}{327}$$

$$\text{From } (ii) \frac{f}{f'} = -\frac{\omega}{\omega'} = -\frac{10/519}{10/327}$$

$$= -\frac{327}{519} = -\frac{109}{173} \qquad \ldots (v)$$

$$\text{From } (i) \qquad \frac{1}{30} = \frac{1}{f} + \frac{1}{f'}$$

$$\text{or} \qquad \frac{f}{30} = 1 + \frac{f}{f'}.$$

Putting the value of $\frac{f}{f'}$ from (v)

$$\frac{f}{30} = 1 - \frac{109}{173} = \frac{173 - 109}{173} = \frac{64}{173}.$$

$$\therefore \qquad f = \frac{64}{173} \times 30 = \frac{1920}{173} = 11.1 \text{ cm.}$$

$$\text{and} \qquad f' = -\frac{173}{109} f = -\frac{173}{109} \times \frac{1920}{173} = -\frac{1920}{109}$$

$$= -17.6 \text{ cm}$$

Since the free surface of the divergent lens is plane, $r'_2 = \infty$.

$$\text{Now,} \qquad \frac{1}{f'} = (\mu' - 1)\left(\frac{1}{r'_1} - \frac{1}{r'_2}\right)$$

Substituting values, we have

$$-\frac{109}{1920} = (1.654-1)\left(\frac{1}{r_1'}-\frac{1}{\infty}\right) = \frac{0.654}{r_1'}$$

$$r'_1 = -\frac{654}{1000} \times \frac{1920}{109} = -\frac{288}{25} = -11.52 \text{ cm.}$$

For the crown glass lens

$$\frac{1}{f} = (\mu-1)\left(\frac{1}{r_1}-\frac{1}{r_2}\right).$$

whence, $f = \frac{1920}{173}$; $\mu = 1.519$; $r_2 = r'_1 = -\frac{288}{25}$

$$\therefore \qquad \frac{173}{1920} = 0.519\left(\frac{1}{r_1}+\frac{25}{288}\right)$$

$$= \frac{519}{1000\,r_1} + \frac{519 \times 25}{1000 \times 288},$$

or

$$\frac{519}{1000\,r_1} = \frac{173}{1920} - \frac{519 \times 25}{288000}$$

$$= \frac{173 \times 150 - 519 \times 25}{288000}$$

$$= \frac{25(1038-519)}{288000} = \frac{25 \times 519}{288000}$$

$$\therefore \qquad r_1 = \frac{519 \times 288}{25 \times 519} = \frac{288}{25} = 11.52 \text{ cm.}$$

Thus all the curved surfaces have the same radii of curvature.

Achromatic Combination of two Co-axial Lenses Separated by a Distance : It is possible to make an achromatic combination of two lenses of the *same material* and of the same sign provided they are separated by a finite distance.

Suppose two lenses of focal lengths f, f' and mean refractive index μ are situated a distance a apart. The equivalent focal length of the combination is given by

$$\frac{1}{F} = \frac{1}{f} + \frac{1}{f'} - \frac{a}{ff'}$$

If F_r and F_b be the focal lengths of the equivalent lenses for red and blue light, then

$$\frac{1}{F_r} = \frac{1}{f_r} + \frac{1}{f_r'} - \frac{a}{f_r f_r'}$$

$$= \frac{\mu_r - 1}{\mu - 1} \cdot \frac{1}{f} + \frac{\mu_r - 1}{\mu - 1} \cdot \frac{1}{f'} - \left(\frac{\mu_r - 1}{\mu - 1}\right)^2 \cdot \frac{a}{ff'}$$

$$= \frac{\mu_r - 1}{\mu - 1}\left(\frac{1}{f} + \frac{1}{f'}\right) - \left(\frac{\mu_r - 1}{\mu - 1}\right)^2 \cdot \frac{a}{ff'} \quad \dots (a)$$

Similarly, $$\frac{1}{F_b} = \frac{\mu_b - 1}{\mu - 1}\left(\frac{1}{f} + \frac{1}{f'}\right) - \left(\frac{\mu_b - 1}{\mu - 1}\right)^2 \cdot \frac{a}{ff'} \quad \dots (b)$$

For achromatism, $F_b = F_r$, that is, left hand sides of (a) and (b) are equal. Subtracting right-hand side of (a) from that of (b), we have

$$\frac{\mu_b - \mu_r}{\mu - 1}\left(\frac{1}{f} + \frac{1}{f'}\right) - \frac{(\mu_b - 1)^2 - (\mu_r - 1)^2}{(\mu - 1)^2} \times \frac{a}{ff'} = 0,$$

or

$$\frac{\mu_b - \mu_r}{\mu - 1}\left(\frac{1}{f} + \frac{1}{f'}\right) - \frac{(\mu_b - \mu_r)(\mu_b + \mu_r - 2)}{(\mu - 1)^2} \times \frac{a}{ff'} = 0,$$

Since $(\mu_b + \mu_r) = 2\mu$, we have

$$\frac{\mu_b - \mu_r}{\mu - 1}\left(\frac{1}{f} + \frac{1}{f'}\right) - \frac{2(\mu_b - \mu_r)(\mu - 1)}{(\mu - 1)^2} \times \frac{a}{ff'} = 0$$

or $$\frac{\mu_b - \mu_r}{\mu - 1}\left(\frac{1}{f}+\frac{1}{f'}\right) - \frac{2a(\mu_b - \mu_r)}{(\mu - 1)\, ff'} = 0$$

or $$\frac{\mu_b - \mu_r}{\mu - 1}\left(\frac{1}{f}+\frac{1}{f'} - \frac{2a}{ff'}\right) = 0$$

or $$\frac{1}{f}+\frac{1}{f'} - \frac{2a}{ff'} = 0$$

$\therefore$ $$\frac{2a}{ff'} = \frac{1}{f}+\frac{1}{f'} = \frac{f+f'}{ff'},$$

or $$2a = f + f'$$

$\therefore$ $$a = \frac{f+f'}{2}. \qquad \text{... (11)}$$

Alternative Method : The reader may prefer the shorter method of calculus.

$$\frac{1}{F} = \frac{1}{f}+\frac{1}{f'} - \frac{a}{ff'}$$

By partial differentiation, we get

$$d\left(\frac{1}{F}\right) = d\left(\frac{1}{f}\right) + d\left(\frac{1}{f'}\right) - ad\left(\frac{1}{ff'}\right)$$

or $$-\frac{dF}{F^2} = -\frac{df}{f^2} - \frac{df'}{f'^2} - a\left[-\frac{df}{f^2 f'} - \frac{df'}{ff'^2}\right],$$

or $$-\frac{dF}{F^2} = \frac{\omega}{f} + \frac{\omega}{f'} - a\left(\frac{w}{ff'} + \frac{\omega}{ff'}\right)$$

Since both the lenses are of the *same material.*

$$\omega = \omega'$$

$\therefore$ $$-\frac{dF}{F^2} = \frac{\omega}{f} + \frac{\omega}{f'} + \frac{2a\omega}{ff'}$$

For achromatism $dF = 0$,

$$\therefore \qquad \frac{\omega}{f} + \frac{\omega}{f'} - \frac{2a\omega}{ff'} = 0$$

$$\text{or} \qquad \frac{1}{f} + \frac{1}{f'} = \frac{2a}{ff'} = 0,$$

$$\text{or} \qquad \frac{2a}{ff'} = \frac{1}{f} + \frac{1}{f'} = \frac{f+f'}{ff'},$$

$$\text{or} \qquad 2a = f + f'$$

$$\text{or} \qquad a = \frac{f+f'}{2}$$

Thus, we get an achromatic combination if the two lenses are separated by a distance equal to half the sum of their focal lengths. This condition is used in the construction of Huygen's eye-piece.

Here we find that ω is eliminated so the result holds good for *all colours, viz.,* when this condition for achromatism is satisfied all the coloured images are equal in size. This is, however, unlike the achromatic doublet already discussed in which the achromatism is true for *two* colours only, on account of irrationality of dispersion in lenses of different materials.

The student, however, should not infer that in this combination, the principal foci for all the colours coincide since red and blue-images occupy different positions and are unequally magnified. It only means that rays of all colours initially parallel to the axis are deviated through the same angle so that each coloured image subtends the same angle at the eye causing the sensation of whiteness due to overlapping. We have, therefore, achromatized *the size* and *not the position* of the coloured images. It is, in fact, impossible to-achromatise a lens system consisting of two single lenses for both size and position of the images. The achromatism of the combination is then similar to that of an ordinary magnifying glass.

The Rainbow

The rainbow is probably the most interesting example of the dispersion of white light and production of solar spectrum in nature. When the sun shines on rain drops during or after a shower they disperse light by refraction and deviate its component colours by internal reflection to the eye of the observer standing with his back turned towards the sun. He sees one or sometimes two rainbows which form arcs of concentric circles with their common centre on the line joining the sun to the eye of the observer.

The inner or the primary bow is the brightest of the two and its radius subtends an angle of 41° at the observer's eye. It exhibits all the colours of the solar spectrum, red being on the outside and violet on the inside. It is formed by the rays which have suffered one internal reflection before finally emerging from the water drops.

The outer bow is called the secondary bow and its radius subtends an angle of about 52° at the observer's eye. It is not so bright as the primary and presents the same succession of colours but in the *reverse* order red on the inner and violet on the outer edge. It is due to the rays which have suffered two internal reflections before coming out of the drops.

In addition to these two principal bows sometimes other bows are also observed near the inner edge of the primary bow and the outer edge of the secondary bow. They are called supernumerary or spurious bows and are due to diffraction of light on emergence from rain drops, their nature depending upon the size of the drops. We also find that the region of the sky lying between the primary and secondary bows is darker while that within the primary and outside the secondary is appreciably brighter than the rest of the sky. Let us now explain these phenomena.

The Primary Rainbow : To understand the formation of the rainbow consider a solar ray SA tailing at an angle of incidence i on a spherical rain drop ABC. Some of the light enters the drop in the direction AB making an angle of refractation equal to r with the normal OA. It meets the outer surface of the drop at B where internal reflection occurs and the ray finally emerges out of the drop at C along CE. From the geometry we find that

$$\angle OAB = \angle OBA = \angle OBC = \angle OCB = r.$$

so that the angle of emergence is equal to the angle of incidence i.

The deviation of the ray at first refraction is $(i - r)$, on reflection at B $(180° - 2r)$ and at second refraction $(i - r)$. Since the deviations are all clockwise, the total deviation

$$D = 2\ (i - r) + (180° - 2r)$$

$$= 180° + 2i - 4r.$$

For normal incidence the deviation is maximum, *i.e.*, $D = 180°$.

For other directions of incidence the deviation will vary between 180° and a minimum value which is approximately 138°.

It will be seen from the graph between the angles of incidence and the corresponding angles of deviation of the ray after suffering two refractions and one internal reflection that deviation changes slowly in the neighbourhood of minimum deviation. Consequently, most of the emergent light will be closely packed and hence *most intense* for angles of incidence corresponding to minimum deviation position. It is obvious, therefore, that in the theory of bows we are most concerned with the rays which suffer the *least* deviation.

For a ray which has suffered one internal reflection, we have

$$D = 180° + 2i - 4r.$$

To find minimum deviation, differentiate D with respect to i and equate it to zero. We, therefore, get

$$\frac{dD}{di} = 2 - 4\frac{dr}{di} = 0$$

Whence
$$\frac{dr}{di} = \frac{1}{2}.$$

But $\sin i = \mu \sin r$, where μ is the refreactive index of water. It differentiation gives

$$\cos i = \mu \cos r \frac{dr}{di}$$

Substituting in it dr/di, we have $2 \cos i = \mu \cos r$,

or
$$4 \cos^2 i = \mu^2 \cos^2 r = \mu^2 (1 - \sin^2 r)$$
$$= \mu^2 - \mu^2 \sin^2 r = \mu^2 - \sin^2 i,$$

or
$$4(1 - \sin^2 i) = \mu^2 - \sin^2 i, \text{ or } 3 \sin^2 i = 4 - \mu^2.$$

$\therefore$
$$\sin i = \sqrt{\frac{4 - \mu^2}{3}}$$

For red light in water,

$$\mu = 1.329,$$

and for violet light in water,

$$\mu = 1.343.$$

This gives $i = 59.6°$, and $r = 40.5°$, so that $D = 137.7°$ for red (approx.) and $i = 59.6°$ and $r = 39.6°$, so that $D = 139.6°$ for violet (approx.). So for any drop illuminated by the sun, the resultant beam of light formed after one internal reflection will be most intense in the direction for which the deviation is 137.7° for red light and 139.6° for violet light.

Since this phenomenon is the same in all planes which pass through the line joining the sun with the observer's eye, the assemblage of drops which lie on the arcs of concentric circles with centres on this line will give rise to the primary rainbow. These circles form the bases of the cones with apex at the observer's eye and the semi-vertical angle of the cones being (180° – 137.7°) = 42.3° for the red light and (180° – 139.6°) = 40.4° for the voilet. Therefore, the primary bow is voilet on the *inner* side and red on the *outer* side, other colours falling in between. The centre of this circle of light will be immediately opposite to the sun and will be below the horizon thus only an are of circle is seen. For an observer on the ground, the sun's altitude must not be more than 42°, if primary bow is to be visible.

The Secondary Rainbow : It can be easily seen that deviation at each internal reflection inside the sphere is (180° - 2r) so the deviation of a ray emerging after *n* internal reflections,

$$D = 2\,(i - r) + n\,(180° - 2r).$$

For certain angles of incidence the light is reflected twice within the rain drops before it emerges in the direction of the observer and it gives rise to the secondary bow.

Putting $n = 2$ in the above relation, the total deviation in this case

$$D = 2(i - r) + 2\,(180° - 2r)$$

$$= 360° + 2i - 6r.$$

Proceeding as in the case of primary rainbow, we find that for minimum deviation, the angle of incidence is given by

$$\text{Sin}\ i = \sqrt{\frac{9 - \mu^2}{9}} \qquad ...(11)$$

Taking the value of μ for red and violet as 1.329 and 1.343 respectively, we obtain the values of the angles of minimum deviation for red light as 231° and for violet 234° approx. We find that the emergent rays in this case cross the incident rays from the sun, the more efficacious rays being packed in the neighbourhood of the surface of the cone of which the semi-vertical angle is (234° — 180°) = 54° for violet and (231° – 180°) = 51° for red. Thus another bow is formed with the order of colours reversed, the violet being on the outside and red on the inside.

This bow is fainter than the primary for two reasons (*i*) the light has undergone two internal reflections and has thereby been weakened and *(ii)* there is greater angular dispersion in this bow than in the primary. These two bows are the only ones, which are usually seen although higher bows with three or four internal reflections exist but as they appear on the same side as the sun, they are lost in its brightness.

The manner in which the observer sees both the primary and secondary bows, the drops between *A* and *B* giving rise to secondary rainbow and those between C and *D*, the primary bow. Drops lying below the drop *D* would send light to the observer after one internal reflection but at a deviation greater than the minimum so that some illumination from the region within this bow will reach the observer. Similarly, light which undergoes two internal reflections from some drops above *A* will give a faint illumination above the secondary bow. In both cases there is an overlapping of all the different colours There is, however, no possibility of reflection of light to the observer *O* from the drops lying in between the two bows, far that would require a deviation less than the minimum for either of the bows This explains why the region of the sky between the primary and the secondary bows is darker and that within the primary and outside the secondary is appreciably brighter than the rest of the sky.

4

The Lenses

Thick Lens

In dealing with thin lenses we assumed that the thickness of a lens is small and hence negligible as compared to the focal length, the object and image distances, which are always measured from the optical centre of the lens. But, when the lens is sufficiently thick this procedure of measuring distances is not applicable and we have to deal with it like a system of coaxial lenses treated as a unit which may or may not be in contact. Such a lens or a system of lenses must be treated as a thick lens.

To find the position of image in this case we may apply the equation of a single refracting surface successively to each surface. But with a large number of surfaces the procedure is laborious and as such, the treatment suggested by C.F. Gauss is followed. Having studied the properties of thick lenses as a whole he has shown that simple formulae as for thin lenses may be applied as usual provided, however, the distances in all cases are measured from two points fixed relatively to the lens called principal points. Besides these two principal points, there are two principal foci and two nodal points as in coaxial lens systems. When the medium on both sides of the thick lens is the same, the nodal points are coincident with the principal points. These six points, already termed as cardinal points, have as well an important bearing on the theory of thick lenses.

Principal Foci and Principal Points : Rays proceeding from the point F_1 emerge parallel to the axis after refraction from the lens and the parallel incident rays are brought to focus at F_2. F_1 and F_2 are called the first and second *principal foci* respectively. The transverse planes through the points of intersection of the incident and the emergent rays are called the first and second *principal planes* and the points H_1 and H_2 where they cut the principal axis are called *principal points.* These points have the same significance as discussed already, with reference to the coaxial system of lenses. The positions of the points F_1, F_2, H_1 and H_2 are fixed for a lens. If the medium is the same on both sides of the lens, the first and second focal lengths ϕ_1 and ϕ_2 will be numerically equal.

Note that the focal lengths are measured from the principal foci to their respective principal points H_1 and H_2 and not to their respective vertices.

In general the principal points and the principal foci are not located symmetrically with respect to the lens but are at different distances from the vertices. The dotted lines indicate the approximate positions of the principal planes in the case of a few lenses of the standard shapes.

Now, we shall discuss the behaviour of a thick lens bounded on both sides by the same (less dense) medium, for instance, a glass lens placed in air.

Thick Lens Formulae : Assuming that the medium on both sides of the lens is air, we shall find the positions of the image, the principal points, the principal foci etc., for a thick lens by the same general method as used for combination of thin liases treating the image formed by the first surface as the object for the second.

Let *PQ*, be an object placed at a distance u from the first surface of a lens of thickness t, refractive index μ, with faces of radii of curvature r_1 and r_2, for refraction at the first surface where v' is the distance of the image from the surface and f_1 is the first focal distance of this surface.

$$\frac{\mu}{v'} - \frac{1}{u} = \frac{\mu - 1}{r_1} = -\frac{1}{f_1}, \quad \text{... (1)}$$

where v' is the distance of the image from the surface and f_1 is the first focal distance of this surface.

Since the image formed by the first surface becomes the object for the second surface, we have for refraction at the second surface $u' = v' - t$, due regard being paid to the sign of v'.

Since the rays are now going from a denser to a rarer medium, we have by the same relation.

$$\frac{\frac{1}{\mu}}{\mu} - \frac{1}{v' - t} = \frac{\frac{1}{\mu} - 1}{r_2} \quad \text{... (2)}$$

or
$$\frac{1}{v} - \frac{\mu}{v' - t} = \frac{1 - \mu}{r_2} = -\frac{\mu - 1}{r_2} = \frac{1}{f_2} \quad \text{... (2a)}$$

where v is the distance of the final image from the second surface and f_2 is the second focal distance for the second surface. It is the distance of the point from the second surface where a beam of parallel rays traversing the more dense medium incident on the second surface converges after emergence from the surface. It is the value of v obtained by putting $v' = \infty$ in equation (2).

By combining relations (1) and (2*a*) and eliminating v', we obtain by algebra,

$$\frac{1}{v-\beta} - \frac{1}{u-\alpha} = \frac{1}{F}$$

where F is the focal length of the thick lens, α and β are the distances of the first principal plane and the second principal plane from the first surface and the second surface of the lens respectively.

Here, $$F = -\frac{\mu f_1 f_2}{\mu(f_2 - f_1) - t} \quad \text{... (4)}$$

$$\beta = -\frac{f_2 t}{\mu(f_2 - f_1) - t} \quad \text{... (5)}$$

and $$\alpha = -\frac{f_1 t}{\mu(f_2 - f_1) - t}$$

If we measure V from a point at a distance β from the second surface and U from a point at a distance α from the first surface, then we may write

$$V = v - \beta$$

and $$U = u - \alpha$$

Equation (3) thus becomes

$$\frac{1}{V} - \frac{1}{U} = \frac{1}{F}, \quad \text{... (7)}$$

a formula identical with that used for a thin lens. U and V are known as the reduced object distance and the reduced image distance respectively.

The point at a distance α from the first surface of the lens is called the first principal point of the lens and the point at a distance β from the second surface of the lens is termed the second principal point. Planes drawn perpendicular to the axis through the principal points are called the *principal planes* of the lens. Knowing the constants of the lens the values of β and α are calculated from equations (5) and (6) and hence the positions of principal points are determined.

Having thus located the principal points, the principal foci can be easily found. *The first principal focus is at a distance F to the left of the first principal point and the second principal focus is at the same distance to the right of the second principal point.* The first and second focal lengths of a lens surrounded by a uniform medium are, therefore, equal in magnitude but opposite in sign.

Since α, β and *F* depend only on the constants r_1, r_2, *t* and μ, their values in any given problem can be readily found. Hence equation (3) provides us with the desired means of locating the image formed by a thick lens.

The positions of the principal points H_1, H_2, the principal foci F_1, F_2 and the course of rays forming the image *P′Q′* of an object *PQ* placed at a distance *u* from the first surface of a thick lens. For tracing the course of rays.

The relations (3) to (6) can be obtained as below :

For refraction at the first and the second surface of a thick lens, we have

$$\frac{\mu}{v'} - \frac{1}{u} = -\frac{1}{f_1} \qquad \text{... } (i)$$

and
$$\frac{1}{v} - \frac{\mu}{v'-t} = \frac{1}{f_2}. \qquad \text{... } (ii)$$

From (*i*),
$$\frac{\mu}{v'} = \frac{1}{u} - \frac{1}{f_1} = \frac{f_1 - u}{uf_1}$$

∴
$$v' = \frac{\mu u f_1}{f_1 - u}. \qquad \text{... } (iii)$$

From (*ii*), $$\frac{\mu}{v'-t} = \frac{1}{v}-\frac{1}{f_2} = \frac{f_2-v}{vf_2}.$$

$$\therefore \qquad v'-t = \frac{\mu v f_2}{f_2-v},$$

or $$v' = t+\frac{\mu v f_2}{f_2-v}, \qquad \ldots (iv)$$

Equating the values of v' in (*iii*) and (*iv*), we have

$$\frac{\mu v f_1}{f_1-u} = t+\frac{\mu v f_2}{f_2-v}$$

Multiplying throughout by $(f_1-u)(f_2-v)$, we get

$$\mu u f_1 (f_2-v) = t(f_2-v)(f_1-u)+\mu v f_2 (f_1-u),$$

or $$\mu u f_1 (f_2-v) - t(f_2-v)(f_1-u) - \mu v f_2 (f_1-u) = 0$$

Opening the brackets and collecting the terms in uv, u and v.

$$\mu u f_1 f_2 - \mu u v f_1 - t f_1 f_2 + t u f_2 + t v f_1 - t u v - \mu v f_1 f_2 + \mu u v f_2 = 0,$$

or $$\mu u v f_2 - \mu u v f_1 - t u v + \mu u f_1 f_2 + t u f_2 - \mu v f_1 f_2 + t v f_1 - t f_2 f_1 = 0,$$

or $$uv[\mu(f_2-f_1)-t] + u[\mu f_1 f_2 + t f_2] - v[\mu f_1 f_2 - t f_1] - t f_1 f_2 = 0$$

Dividing throughout by $\mu(f_2-f_1)-t$, we have

$$uv+\frac{u(\mu f_1 f_2 + t f_2)}{\mu(f_2-f_1)-t}-\frac{v(\mu f_1 f_2 + t f_2)}{\mu(f_2-f_1)-t}-\frac{t f_1 f_2}{\mu(f_2-f_1)-t} = 0. \ldots (v)$$

the equation can be put in the form

$$\frac{1}{v-\beta}-\frac{1}{u-\alpha} = \frac{1}{F} \qquad \ldots (vi)$$

Multiplying (*vi*) throughout by $F(u-\alpha)(v-\beta)$ and collecting terms in uv, u and v, we obtain

$$uv - u(F + \beta) + v(F - \alpha) - F(\beta - \alpha) + \alpha\beta = 0 \quad \dots (vii)$$

In order that (*v*) and (*vii*) should be identical , we must have

$$-F - \beta = \frac{\mu f_1 f_2 + t f_2}{\mu(f_2 - f_1) - t} \quad \dots (viii)$$

$$-F + \alpha = \frac{\mu f_1 f_2 + t f_1}{\mu(f_2 - f_1) - t} \quad \dots (ix)$$

and $F(\beta - \alpha) - \alpha\beta = \dfrac{t f_1 f_2}{\mu(f_2 - f_1) - t}$ ... (*x*)

Thus (*viii*), (*ix*) and (*x*) are three simultaneous equation to find the three unknown α, β and F.

To solve these, first multiply the corresponding sides of (*viii*) and (*ix*). This gives us

$$F^2 - F\alpha + F\beta - \alpha\beta = \frac{\mu f_1 f_2 + t f_2}{\mu(f_2 - f_1) - t} + \frac{\mu f_1 f_2 + t f_1}{\mu(f_2 - f_1) - t} \quad \dots (xi)$$

Subtracting (*x*) from (*xi*), we get

$$F^2 = \frac{(\mu f_1 f_2)^2 - \mu f_1^2 f_2 t + \mu f_1 f_2^2 t - t^2 f_1 f_2}{[\mu(f_2 - f_1) - t]^2} - \frac{t f_1 f_2}{\mu(f_2 - f_1) - t}$$

$$= \frac{(\mu f_1 f_2)^2 - \mu f_1^2 f_2 t + \mu f_1 f_2^2 t - t^2 f_1 f_2 - t f_1 f_2[\mu(f_2 - f_1) - t]}{[\mu(f_2 - f_1) - t]^2}$$

$$= \frac{(\mu f_1 f_2)^2 + t f_1 f_2[\mu(f_2 - f_1) - t] - t f_1 f_2[\mu(f_2 - f_1) - t]}{[\mu(f_2 - f_1) - t]^2}$$

$$= \frac{(\mu f_1 f_2)^2}{[\mu(f_2 - f_1) - t]^2}$$

$$\therefore \quad F = \pm \frac{\mu f_1 f_2}{\mu(f_2 - f_1) - t} \quad \dots (xii)$$

In order to find out whether the positive or the negative sign is to be retained it should be noticed that it must be true for all values of t. For a thin lens $t = 0$, and obviously f_1 and f_2 are conjugate focal distances and α and β are both zero.

Substituting $\alpha = 0$, $u = f_1$ and $\beta = 0$, $v = f_2$ in (*vi*), we have

$$\frac{1}{f_2} - \frac{1}{f_1} = \frac{1}{F}$$

$$\therefore \qquad F = -\frac{f_1 f_2}{f_2 - f_1}.$$

Substituting $t = 0$ in (*xii*), we see that we obtain value of F just found, provided that we take the negative sign. Hence

$$F = \frac{\mu f_1 f_2}{\mu(f_2 - f_1) - t} \qquad \text{... } (xiii)$$

Adding (*xiii*) and (*viii*), we have

$$-\beta = \frac{\mu f_1 f_2 + t f_2}{\mu(f_2 - f_1) - t} - \frac{\mu f_1 f_2}{\mu(f_2 - f_1) - t}$$

$$= \frac{t f_2}{\mu(f_2 - f_1) - t}$$

or

$$\beta = \frac{f_2 t}{\mu(f_2 - f_1) - t} \qquad \text{... } (xiv)$$

Adding (*xiii*) and (*ix*), we get

$$\alpha = \frac{\mu f_1 f_2 - t f_1}{\mu(f_2 - f_1) - t} - \frac{\mu f_1 f_2}{\mu(f_2 - f_1) - t}$$

$$= -\frac{f_1 t}{\mu(f_2 - f_1) - t} \qquad \text{... } (xv)$$

Hence

$$\alpha f_2 = \beta f_1, \qquad \text{... } (xvi)$$

an important relation which is useful in the theory of thick lenses.

Power of a Thick Lens : The power of a thick lens is given by

$$P = \frac{1}{F} = -\frac{\mu(f_2 - f_1)t}{\mu f_1 f_2}$$

$$= -\frac{1}{f_1} + \frac{1}{f_2} + \frac{t}{\mu f_1 f_2}$$

$$= \frac{\mu - 1}{r_1} - \frac{\mu - 1}{r_2} - \frac{t}{\mu}\left(\frac{\mu - 1}{r_1}\right)\left(-\frac{\mu - 1}{r_2}\right)$$

$$= (\mu - 1)\left(\frac{1}{r_1} - \frac{1}{r_2}\right) + \frac{t(\mu - 1)^2}{\mu r_1 r_2} \qquad \text{... (8)}$$

The power of a thick lens is also obviously given by

$$P = P_1 + P_2 - \frac{t}{\mu} P_1 P_2,$$

where P_1 and P_2 are the powers of the two surfaces respectively.

Graphical Construction of the Image of an Object on the Axis of a Thick Lens : If F be the focal length of a thick lens and the object and the image distances be measured from the first and second principal planes respectively, the usual graphical method of locating the images for thin lenses will apply to thick lenses as well.

PQ is an object, the position of whose image we wish to find graphically. First calculate from the given data the values of F, α and β. Then locate on the axis the principal points H_1 and H_2 by measuring from the first vertex $O_1H_1 = \alpha$, to the right if α is positive and to the left if it is negative, and from the second vertex $O_2H_2 = \beta$ in the same way while the principal planes are drawn though H_1 and H_2. The points F_1 and F_2, the first and the second principal foci are then located by taking $H_1F_1 = -F$ *and* $H_2 F_2 = -F$. For the purposes of graphical construction of the image, the lens may be regarded as-replaced by its two principal planes and the properties of these planes mentioned already with reference to coaxial systems are taken into consideration.

It will be seen that ray 1 from P drawn parallel to the optic axis meets the first principal plane at M and the second at N at the same height and then passes through F_2, the second principal focus of the lens.

Similarly, ray 2 passing through F_1, the first principal focus of the lens meets the first principal plane at M' and is then rendered parallel to the axis. The point of intersection of these two rays gives, the required position of P', the image of P.

From P' drop $P'Q'$ perpendicular to the axis meeting it in Q, the $P'Q'$ is the image of PQ.

It may be noted that here all the rays in the region between two principal planes are drawn parallel to the axis since by definition these planes are conjugate ones, for which the magnification is unity and has a positive sign. It means that a small object on the axis in one of these planes has an equal-sized, erect image in the other. These planes are, therefore, known as *unit planes*.

Example : *Calculate the positions of the principal points and principal foci of a thick double convex lens, the radii of whose surfaces are 4 cm and 3 cm, the thickness of the lens being 2.5 cm and the refractive index 1.50.*

Here, $r_1 = +4$ cm; $r_2 = -3$ cm, $t = 2.5$ cm, $\mu = 1.5$.

$$\therefore \qquad f_1 = -\frac{r_1}{\mu - 1} = -\frac{+4}{1.5 - 1} = -8 \text{ cm.}$$

$$f_2 = -\frac{r_2}{\mu - 1} = -\frac{-3}{1.5 - 1} = +6 \text{ cm}$$

$$F = -\frac{\mu f_1 f_2}{\mu(f_2 - f_1) - t} = -\frac{1.5 \times (-8) \times (+6)}{1.5[6-(-8)] - 2.5}$$

$$= +\frac{1.5 \times 48}{18.5} = +\frac{144}{37} = +\mathbf{3.89\ cm}\text{ approx.}$$

$$\alpha = -\frac{f_1 t}{\mu(f_2 - f_1) - t}$$

$$= \frac{-8 \times 2.5}{1.5(6+8) - 2.5} = +1.08\text{ cm.}$$

$$\beta = \frac{f_2 t}{\mu(f_2 - f_1) - t}$$

$$= -\frac{+6 \times 2.5}{1.5(6+8) - 2.5} = -\frac{15}{18.5} = -0.81\text{ cm;}$$

$\therefore$ $O_1H_1 = \alpha = +1.08$ cm;

$O_2H_2 = \beta = -0.81$ cm;

$H_1F_1 = F = 3.89$ cm;

$\therefore$ $O_1F_1 = H_1F_1 - O_1H_1$

$= 3.89 - 1.08 =$ **2.81 cm.**

$H_2F_2 = F = 3.89$ cm;

$\therefore$ $O_2F_2 = H_2F_2 - O_2H_2$

$= 3.89 - 0.81 =$ **3.08 cm.**

Example : *With the following data for a thick lens, calculate f_1, f_2, F, α and β and also the distances of the principal points and principal foci from the first vertex of the lens.*

$r_1 = +3$ cm; $t = 2$ cm; $r_2 = -5$ cm; $\mu = 1.50$.

Here, $$f_1 = -\frac{r_1}{\mu - 1} = -\frac{+3}{1.5 - 1}$$

$$= -6\text{ cm.}$$

$$f_2 = -\frac{r_2}{\mu - 1} = -\frac{-5}{1.5 - 1} = +\ 10 \text{ cm.}$$

$$F = -\frac{\mu f_1 f_2}{\mu\left(f_2 - f_1\right) - t}$$

$$= -\frac{1.50 \times (-6) \times (+10)}{1.50\left[+10 - (-6) - 2\right]} = 4.09 \text{ cm.}$$

$$\beta = -\frac{f_2 t}{\mu\left(f_2 - f_1\right) - t}$$

$$= -\frac{(+10) \times (+2)}{1.50\left(+10 + 6\right) - 2} = -\ 0.90 \text{ cm.}$$

$$\alpha = -\frac{f_1 t}{\mu\left(f_2 - f_1\right) - t}$$

$$= -\frac{(-6) \times (+2)}{1.5\left(+10 + 6\right) - 2} = +\ 0.54 \text{ cm.}$$

The distance of the first principal point from the first vertex, $\alpha = +\ 0.54$ cm, that is towards its right.

The distance of the second principal point from the second vertex = – 0.91 cm, that is, towards its left.

$\therefore$ its distance from the first vertex = 2 – 0.91 = 1.09 cm.

The distance of the first principal focus from the first vertex = 4.09 – 0.54 = 3.55 cm and the distance of the second principal focus from the first vertex = 2 + (4.09 – 0.91) = 5.18 cm.

Example : *An object is placed 50 cm from a thick double convex lens, with faces of radii of curvature 15 cm, and 20 cm, thickness 15 cm and refractive index 1.50. Find the position of the image.*

Here, we have

$r_1 = +\,15$ cm; $r_2 = -\,20$ cm; $t = 1.5$ cm; $\mu = 1.50$, $u = -\,50$ cm

$$\therefore \quad f_1 = -\frac{r_1}{\mu - 1} = -\frac{+15}{1.50 - 1} = -\,30.0 \text{ cm.}$$

$$f_2 = -\frac{r_2}{\mu - 1} = -\frac{-20}{1.50 - 1} = +\,40 \text{ cm.}$$

$$F = -\frac{\mu f_1 f_2}{\mu (f_2 - f_1) - t}$$

$$= -\frac{1.50 \times (-30) \times (+40)}{1.50\,[+40 + 30] - 1.5} = +\,17.4 \text{ cm.}$$

$$\alpha = -\frac{f_1 t}{\mu (f_2 - f_1) - t}$$

$$= -\frac{(-30) \times (+1.5)}{1.50\,(+40 + 30) - 1.5} = +\,0.43 \text{ cm.}$$

$$\beta = -\frac{f_2 t}{\mu (f_2 - f_1) - t}$$

$$= -\frac{(+40) \times (+1.5)}{1.50\,(+40 + 30) - 1.5} = -\,0.58 \text{ cm.}$$

Applying the relation,

$$\frac{1}{v - \beta} - \frac{1}{u - \alpha} = \frac{1}{F}$$

and substituting the values of α, β and F, we have

$$\frac{1}{v - (-\,0.58)} - \frac{1}{-\,50 - (+\,0.43)} = +\frac{1}{17.4}$$

o $\quad r \dfrac{1}{v + 0.58} - \dfrac{1}{50.43} = +\dfrac{1}{17.4}$,

where $v = 26.0$ cm.

Refraction by a Thick Lens Separating Media of Different Refractive Indices : Let us suppose that the refractive index of the material of the lens is μ_2, that of the medium on the left μ_1 and that of the medium on the right μ_2.

Then for the first refraction,

$$\frac{\frac{\mu_2}{\mu_1}}{v'} - \frac{1}{u} = -\frac{1}{f_1}, \qquad \text{... (a)}$$

where $$\frac{1}{f_1} = -\frac{\mu_2 - \mu_1}{\mu_1}\frac{1}{r_1}.$$

For the second refraction,

$$\frac{1}{v} - \frac{\frac{\mu_2}{\mu_3}}{v' - t} = \frac{1}{f_2} \qquad \text{... (b)}$$

where $$\frac{1}{f_2} = \frac{\mu_3 - \mu_2}{\mu_3} \times \frac{1}{r_2}. \qquad \text{... (b')}$$

Changing (*a*) and (*b*) into the forms similar to (*i*) and (*ii*), we get

$$\frac{\mu_2}{v'} - \frac{1}{\frac{u}{\mu_1}} = -\frac{1}{\frac{f_1}{\mu_1}} \qquad \text{... (c)}$$

and $$\frac{1}{\frac{v}{\mu_2}} - \frac{\mu_2}{v' - t} = \frac{1}{\frac{f_2}{\mu_3}}. \qquad \text{... (d)}$$

The terms $\mu_2, u/\mu_1, v/\mu_3, f_1/\mu_1$ and f_2/μ_3 of these equations correspond to the terms μ, u, v, f_1 and f_2 of equations (*i*) and (*ii*).

The solution of these equations is obtained by substituting: these values in equations (*vi*), (*xiii*), (*xiv*) and (*xv*).

Equation (*vi*) thus becomes

$$\frac{1}{\frac{v}{\mu_2}-\beta}-\frac{1}{\frac{u}{\mu_1}-\alpha}=\frac{1}{F},$$

or $$\frac{\mu_2}{v-\mu_3\beta}-\frac{\mu_1}{u-\mu_1\alpha}=\frac{1}{F} \quad \text{... } (e)$$

Putting $U = u - \mu_1\alpha$ and $V = v - \mu_2\beta$, we may rewrite this equations as

$$\frac{\mu_3}{V}-\frac{\mu_1}{U}=\frac{1}{F} \quad \text{... } (f)$$

From Eqn. (*f*) it is obvious that keeping in mind the definitions of first and second focal lengths, the first focal length (ϕ_1) of the lens will be equal to $-\mu_1 F$ and the second focal length (ϕ_2) will be equal to $\mu_3 F$.

Then $$\frac{\mu_3}{V}-\frac{\mu_1}{U}=-\frac{\mu_1}{\phi_1}=\frac{\mu_3}{\phi_3} \quad \text{... } (g)$$

We also find that the two focal lengths of the lens are *numerically* unequal being in the ratio of the refractive indices of the first and second medium.

Since $$\frac{\phi_1}{\phi_2}=\frac{-\mu_1 F}{\mu_3 F}=-\frac{\mu_1}{\mu_3} \quad \text{... } (h)$$

The respective distance of the first and the second principal point from the first and the second surface of the lens will be equal to

$$\alpha' = \mu_1\alpha$$

and $$\beta' = \mu_2\beta.$$

Writing f_1/μ_1 for f_1, f_2/μ_3 for f_2 and μ in the values of α and β, we have

$$\alpha' = \mu_1 \alpha = -\frac{\mu_1 \dfrac{f_1 t}{\mu_1}}{\mu_2 \left(\dfrac{f_2}{\mu_2} - \dfrac{f_1}{\mu_1} \right) - t}$$

$$= -\frac{\mu_1 \mu_3 f_1 t}{\mu_2 (\mu_2 - \mu_2 f_1) - \mu_1 \mu_2 t} \quad \text{... (2)}$$

Similarly, $$\beta' = \mu_2 \beta = -\frac{\mu_1 \mu_3 f_2 t}{\mu_2 (\mu_2 f_2 - \mu_3 f_1) - \mu_1 \mu_3 t} \quad \text{... (10)}$$

Obviously, $$f_2 \alpha' = f_1 \beta' \quad \text{... (11)}$$

an important relation which is useful in the theory of thick lenses.

Also $$F = -\frac{\mu_2 \dfrac{f_1}{\mu_1} \times \dfrac{f_2}{\mu_3}}{\mu_2 \left(\dfrac{f_2}{\mu_3} - \dfrac{f_1}{\mu_1} \right) - t}$$

$$= -\frac{\mu_2 f_1 f_2}{\mu_2 (\mu_1 f_2 - \mu_3 f_1) - \mu_1 \mu_3 t} \quad \text{... (12)}$$

$\therefore$ $$\phi_2 = \mu_3 F$$

$$= -\frac{\mu_3 \mu_2 f_1 f_2}{\mu_2 (\mu_1 f_2 - \mu_2 f_1) - \mu_1 \mu_3 t} \quad \text{... (12a)}$$

and $$\phi_1 = \mu_1 F$$

$$= -\frac{\mu_1 \mu_2 f_1 f_2}{\mu_2 (\mu_1 f_2 - \mu_3 f_1) - \mu_1 \mu_3 t} \quad \text{... (12b)}$$

With the help of equations (9) and (10) we can locate the positions of principal points. Since in this case the principal points do not coincide with the nodal points, the latter pair of points can be determined by combining with these results equations (16) and (16 *a*).

Having thus located the principal points, the principal foci can be easily found. The first principal focus is at a distance ϕ_1 to the left of the first principal point H_1, and the second principal focus is at a distance ϕ_2 to the right of the second principal point H_2.

It may be noted that the thin lens formulae, $\frac{1}{v} - \frac{1}{u} = \frac{1}{f}$

$$\frac{\mu_3}{v} - \frac{\mu_1}{u} = -\frac{\mu_1}{f_1} = \frac{\mu_3}{f_2}$$

and
$$\frac{f_1}{f_2} = \frac{\mu_1}{\mu_3},$$

connecting the conjugate points are applicable to a thick lens provided that the focal lengths, object and image distances are measured from the principal points and not from the vertices as in a thin lens.

The Principal Points of a Lens are Conjugate Foci : This can be proved in the case of a lens separating media of different refractive indices and then it evidently follows that it applies to a lens under any conditions.

Putting β' for $\mu_2\beta$ and α' for $\mu_1\alpha$ in equation *(e)*, we have

$$\frac{\mu_2}{v - \beta'} - \frac{\mu_1}{u - \alpha'} = \frac{1}{F}$$

Simplifying, $$v - \beta' = \frac{\mu_2 F \times (u - \alpha')}{\mu_1 F + (u - \alpha')} \qquad ...(13)$$

When $u = \alpha$, we get $v - \beta' = 0$ or $v = \beta'$ for all values of F, μ_3, μ_1.

Hence an object at a distance α' from the first surface of the lens (i e., an object at the first principal point) will produce an image at β′ from the second surface of the lens (i.e., at the second principal point).

The Principal Planes of a Thick Lens are Planes of Unit Magnification : Let an object PQ, be placed on the axis of the lens and let F_1 be the first focal point of the first surface of the lens so that a ray of light PA starting from P and appearing to come from F_1 is rendered parallel to the axis along AB after refraction at the first surface. After refraction at the second surface the same ray will pass through F_2 the second focal point of the second surface of the lens. From the relations obtained already the position of the image P_1Q_1 can be located, that of P_1 being corresponding to the object point P. Drop perpendiculars AC and BD on the axis. When the slopes of the rays F_1A and BF_2 are small the feet of the perpendiculars will almost coincide with O_1 and O_3 the intersections of the axis with the faces of the lens.

Let $\quad AC = BD = h. \quad O_1Q = u$

and $O_2Q_1 = v$, so that

$$CQ = O_1Q = u \text{ approx.}$$

and $$DQ_1 = O_2Q_1 \doteq v \text{ approx.}$$

If f_1 is the first focal distance of the first surface $(= - CF_1)$ and f_2 the second focal distance of the second surface $(= DF_2)$ and if the heights of the object and the image are h_1 and h_2 respectively, then by similar Δs ACF_1 and PQF_1,

$$\frac{AC}{CF_1} = \frac{PQ}{QF_1} = \frac{PQ}{CF_1 - CQ},$$

or *conventionally*, $\dfrac{h}{-f_1} = \dfrac{h_1}{-f_1-(-u)} = \dfrac{h_1}{u-f_1}$

Also from Δs $P_1Q_1F_2$ and BDF_2

$$\frac{BD}{BF_2} = \frac{P_1O_1}{F_2Q_1} = \frac{P_1O_1}{DQ_1 - DF_2},$$

or *conventionally*, $\frac{h}{f_2} = \frac{-h_2}{v - f_2} = \frac{h_2}{f_2 - v}$.

Let m be the magnification produced by the lens, then

$$\mu = \frac{h_2}{h_1} = \frac{h(f_2 - v)}{f_2} \times \frac{f_1}{h(f_1 - u)}$$

$$= \frac{f_1(f_2 - v)}{f_2(f_1 - u)} = \frac{f_1 f_2 - f_1 v}{f_2 f_1 - f_2 u} \qquad \text{... (14)}$$

If now we suppose that the object lies in the first principal plane of the lens, *i.e.*, $u = \alpha'$ as already proved, the corresponding image will be in the second principal plane, so that $v = \beta'$, hence

$$m = \frac{f_1 f_2 - f_1 \beta'}{f_1 f_2 - f_2 \alpha'} = +1 \qquad \text{... (15)}$$

since $f_2 \alpha' = f_1 \beta'$ by equation (11)

Consequently, any ray directed towards a point in the first principal plane at a distance h from the axis will give rise to a transmitted ray proceeding from a point in the second principal plane on the same side of the axis and at a distance h from it.

Nodal Points : Similar to the case of coaxial systems of lenses there is a pair of conjugate points on the axis of a thick lens called the first and second nodal points. They have the same properties as described earlier in the coaxial systems, *viz.*, that in the general case of a thick lens bounded by two different media *nodal points are defined as points on the axis such that an incident ray directed towards the first nodal point on emerging from the lens proceeds from the second nodal point in a direction parallel to that of the incident ray.* The nodal points and the principal points of a thick lens are also *coincident* when the medium on both sides of the lens is the *same.*

We shall now bring out the different significance of the nodal points and the principal points by considering a thick lens separating two media of different refractive indices in which case the two pairs of points are separate.

Let H_1. H_2 and N_1, N_2, be the principal and the nodal points respectively of a thick lens and AF_1 an object in the first focal plane .of the lens.

Then the ray AN_1 emerges from the lens in the parallel direction N_2D. Let the incident and the emergent rays intersect the principal planes in B and C respectively. Then since principal planes are planes with unit magnification,

$$BH_1 = CH_2$$

and as AN_1 and N_2D are parallel,

$$N_1H_1 = N_2H_2.$$

$\therefore$ obviously $\qquad N_1N_2 + H_1\ H_2. \qquad \ldots (16)$

Thus the distance between the nodal points is exactly equal to the distance between the principal points.

Now draw a ray AE parallel to the axis to cut the first principal plane in E. It will after refraction by the lens proceed from G a point in the second principal plane at an equal distance from the axis and pass through the second principal focus F_2. As A lies in the first focal plane of the lens, all rays proceeding from it will be rendered parallel after passing through the lens. Hence GF_2 is parallel to N_2D and also to AN_1.

Thus the two Δs GF_2H_2 and AN_1F_1 are congruent and hence

$$H_2F_2 = N_1F_1 = H_1F_1 - H_1N_1,$$

or $\qquad H_1N_1 = H_1F_1 = H_2F_2$

Putting conventional signs

$$-H_1N_1 = -H_1F_1 - H_2F_2,$$

or $$H_1\,N_1 = H_1\,F_1 + H_2\,F_2 = \phi_1 + \phi_2, \quad \ldots (16a)$$

where ϕ_1 and ϕ_2 are the first and second focal lengths of the lens.

If the lens is surrounded with a *uniform medium*

$$\phi_1 = -\phi_2$$

and so $$H_1\,N_1 = 0$$

or the nodal points N_1 and N_2 coincide with the principal points H_1 and H_2 of the lens.

In other cases the positions of the nodal points are graphically determined by reversing the process above.

It may be noted that the knowledge of the six cardinal points namely, the two principal foci, the two principal points and the two nodal points is sufficient to solve all problems relating to a thick lens.

Optical Centre of a Lens

Let C_1 and C_2, be centres of curvature the of faces of a lens placed in a uniform medium and draw two parallel radii C_1Q, $C_2\,R$, then the tangents at Q and R will obviously be parallel to each other. Hence a ray PQ which on refraction traverses the path QR will emerge parallel to PQ as the two refractions which it suffered at Q and R occurred at opposite parallel surfaces of a refracting medium. The point O where the ray QR cuts the axis is called the optical centre of the lens.

It is required to locate the position of O. By similar triangles C_1OQ and C_2OR, we have

$$\frac{OC_1}{OC_2} = \frac{QC_1}{RC_2} = \frac{O_1C_1}{O_2C_2} = \frac{r_1}{r_2}.$$

Also since $\dfrac{O_1C_1}{O_2C_2} = \dfrac{OC_1}{OC_2}$

$$= \frac{O_1C_1 - OC_1}{O_2C_2 - OC_2} = \frac{O_1O}{O_2O}$$

$$O_1O : O_2O = r_1 : r_2. \qquad \text{... (17)}$$

That is, for a given lens, the position of O is fixed. Hence any ray which emerges from the lens in a direction parallel to the incident ray passes through point O.

It is through this point that an undeviated ray is drawn in all cases. Next produce the rays PQ and RS forwards and backwards respectively to cut the axis in N_1, N_2, Then N_1, N_2 are the first and the second nodal points of the lens. They are conjugate to O with regard to the *first* and *second* retracting surfaces as a ray incident through N_1 passes through O after *first* refraction and on emergence passes through N_2 parallel to its original direction.

Hence for a lens in a uniform medium the principal points (which coincided in this case with the nodal points) are the points conjugate to the optical centre with regard to the two refracting surfaces. By making use of equation (17), we can locate the positions of the nodal points.

Since $\dfrac{O_1O}{O_2O} = \dfrac{r_2}{r_1}$

and $O_1O_2 = t$, we get

$$= O_2O - t = \frac{r_2}{r_1} O_1O - t,$$

i.e., $O_1O = -\dfrac{r_1 t}{r_2 - r_1}$

If N_1 is regarded as the point object in air for the first surface

$$\frac{\mu}{v'} - \frac{1}{O_1N_1} = \frac{\mu - 1}{r_1},$$

where $v' = -O_1O$.

$$\therefore \quad \frac{\mu(r_2 - r_1)}{r_1 t} - \frac{1}{O_1N_1} = \frac{\mu - 1}{r_1},$$

or

$$= \frac{\mu(r_2 - r_1)}{r_1 t} - \frac{\mu - 1}{r_1}$$

$$= \frac{\mu(r_1 - r_2)}{r_1 t} - \frac{\mu - 1}{r_1}$$

$$= \frac{\mu(r_1 - r_2) - t(\mu - 1)}{r_1 t}$$

$$= \frac{\mu(r_1 - r_2) + t(1 - \mu)}{r_1 t}$$

$$\therefore \quad O_1N_1 = \frac{r_1 t}{\mu(r_1 - r_2) + t(1 - \mu)} \qquad \dots (18)$$

Similarly,

$$O_2N_2 = \frac{r_2 t}{\mu(r_1 - r_2) + t(1 - \mu)} \qquad \dots (19)$$

Positions of cardinal points in some special cases.

Sphere Surrounded by Air.

(i) *Distances fo Principal Points :* The distance of the first principal point from the first vertex,

$$\alpha = O_1H_1 = \frac{f_1 t}{\mu(f_2 - f_1) - t}.$$

Now,

$$f_1 = -\frac{r_1}{\mu - 1} = -\frac{r}{\mu - 1} \qquad \text{since } r_1 = r$$

$$f_2 = -\frac{r_2}{\mu - 1} = \frac{r}{\mu - 1} \qquad \text{since } r_2 = -r$$

and $t = 2r$.

$$\therefore \qquad \alpha = \frac{-\dfrac{r}{\mu-1}\times 2r}{\mu\left(\dfrac{r}{\mu-1}+\dfrac{r}{\mu-1}\right)-2r} = \frac{\dfrac{2r^2}{\mu-1}}{\mu\dfrac{2r}{\mu-1}-2r}$$

$$= \frac{2r^2}{\mu-1}\times\frac{\mu-1}{2\mu r-2\mu r+2r} = \frac{2r^2}{2r} = r.$$

The distance of the second principal point from the second vertex,

$$\beta = O_2H_2 = \frac{f_2 t}{\mu\left(f_2-f_1\right)-t}$$

$$= -\frac{\dfrac{r}{\mu-1}\times 2r}{\mu\left(\dfrac{r}{\mu-1}+\dfrac{r}{\mu-1}\right)\times 2r} = \frac{\dfrac{2r^2}{\mu-1}}{\dfrac{2\mu r}{\mu-1}-2r}$$

$$= -\frac{2r^2}{2r} = -r.$$

Thus, the two principal points of a sphere are coincident with the centre of the sphere. Since the sphere is surrounded with air, the nodal points also coincide with the centre and so is the case with the optical centre. Hence all the points are coincident with the centre of the sphere.

(ii) *Focal Lengths* : The distance of the first principal focus from the first principal plane and that of the second principal focus from the second principal plane, *viz.*, H_1F_1 and H_2F_2 are each equal to *conventionally* $H_1\ F_1$ is negative and H_2F_2 is positive.

$$\text{Now,} \qquad F = -\frac{\mu f_1 f_2}{\mu\left(f_2-f_1\right)-t}$$

$$= -\frac{\mu \times \frac{-r}{\mu-1} \times \frac{r}{\mu-1}}{\mu\left(\frac{r}{\mu-1}+\frac{r}{\mu-1}\right)-2r} = \frac{\frac{\mu r^2}{(\mu-1)^2}}{\mu\left(\frac{2r}{\mu-1}\right)-2r}$$

$$= \frac{\mu r^2}{(\mu-1)^2} \times \frac{\mu-1}{2\mu r - 2\mu r + 2r}$$

$$= \frac{\mu r^2}{2r(\mu-1)} = \frac{\mu r}{2(\mu-1)}$$

The sphere acts as a converging refracting system and the equivalent thin lens would be a convergent lens of numerical focal length $\frac{\mu r}{2(\mu-1)}$ placed at the centre of the sphere.

Plano-convex lens surrounded by air

(i) Distances of principal points.

Here, $r_1 = \infty; \quad r_2 = -r,$

hence $\frac{1}{f_1} = -\frac{\mu-1}{r_1} = \frac{\mu-1}{\infty} = 0$

and $\frac{1}{f_2} = -\frac{\mu-1}{r_2} = \frac{\mu-1}{r}.$

Now, $\alpha = O_1H_1 = -\frac{f_1 t}{\mu(f_2 - f_1) - t}$

$$= \frac{-t}{\mu\left(\frac{f_2}{f_1}-1\right)-\frac{t}{f_1}} = \frac{-t}{\mu\frac{f_2}{f_1}-\mu-\frac{t}{f_1}}$$

$$= +\frac{t}{\mu} \qquad \ldots \left[\text{since } \frac{1}{f_1} = 0\right]$$

and $$\beta = O_2H_2 = -\frac{f_2 t}{\mu(f_2 - f_1) - t}$$

$$= \frac{-t/f_1}{\frac{\mu}{f_1} - \frac{\mu}{f_2} - \frac{t}{f_1 f_2}} = 0\ldots. \left[\text{since } \frac{1}{f_1} = 0\right]$$

As the lens is surrounded by air, the principal points and the nodal points are coincident.

(ii) Focal lengths

$$H_2F_2 = -H_1F_1 = F$$

Now, $$F = -\frac{\mu f_1 f_2}{\mu(f_2 - f_1) - t}$$

$$= -\frac{\mu}{\frac{\mu}{f_1} - \frac{\mu}{f_2} - \frac{t}{f_1 f_2}}$$

$$= -\frac{\mu}{\frac{-\mu}{f_2}} = f_2 \qquad \ldots \left[\text{since } \frac{1}{f_1} = 0\right]$$

or $$F = f_2 = \frac{r}{\mu - 1}$$

If convex side faces the incident beam, $\alpha = 0$,

$$\beta = -\frac{t}{\mu}$$

and $$F = \frac{r}{\mu - 1}$$

Note that in the case of a hemispherical lens, $t = r$,

$$\alpha = \frac{r}{\mu}, \ \beta = 0 \text{ and } F = \frac{r}{\mu - 1}.$$

This means that for a hemispherical lens the distances of the principal foci form the principal points are the same as in the case of a plano-convex lens. If for such a lens $\mu = 3/2$ and PQ is the position of the object, then for refraction at first surface

$$O_1Q_1 = \mu O_1Q = \frac{3}{2}O_1Q$$

$$\alpha = O_1H_1 = \frac{r}{\mu} = \frac{r}{3/2} = \frac{2}{3}r; \quad \beta = O_2H_2 = 0,$$

i.e., H_2 coincides with O_2,

$$F = \frac{r}{\mu - 1} = \frac{r}{3/2 - 1} = \frac{r}{1/2} = 2r,$$

$$\therefore \qquad H_1F_1 = -2r.$$

Now,
$$O_1F_1 = H_1F_1 - H_1O_1$$

$$= H_1F_1 + O_1H_1 = -2r + \frac{2}{3}r = -\frac{4}{3}r,$$

that is, the first principal focus is at distance $4r/3$ to the left of O.

Also $\quad H_2\,F_2 = O_2F_2$, since H_2 and F_2 are coincident.

Since $\quad H_2\,F_3 = F = 2r$,

$\therefore \quad O_2F_2 = 2r.$

Example : *A glass sphere* (μ = 1.50) *has a radius 1.5 cm. Calculate the positions of the cardinal points and also the position of the image when the object is 1.5 cm from the nearest point of the sphere assuming paraxial rays.*

Here, $\quad r_1 = +1.5$ cm, $\quad r_2 = -1.5$ cm,
$t = 3$ cm; $\quad u = 1.5$ cm.

$$\therefore \qquad f_1 = -\frac{r_1}{\mu - 1} = -\frac{+1.5}{1.5 - 1} = -3 \text{ cm}.$$

$$f_2 = -\frac{r_2}{\mu - 1} = -\frac{-1.5}{1.5 - 1} = +3 \text{ cm.}$$

$$\alpha = O_1H_1 = -\frac{f_1 t}{\mu(f_2 - f_1) - t} = -\frac{-3 \times 3}{1.5(3+3) - 3}$$

$= + 9/6 = +$ **1.5 cm.**

$$\beta = O_2H_2 = -\frac{f_2 t}{\mu(f_2 - f_1) - t} = -\frac{3 \times 3}{1.5(3+3) - 3}$$

$= -9/6 = -1.5$ cm.

Thus, the principal points *coincide* with the centre of the sphere. Since the sphere is surrounded with air, the nodal points also coincide with the centre and so is the case with the optical centre.

$$F = H_2F_2 = -\frac{\mu f_1 f_2}{\mu(f_2 - f_1) - t}$$

$$= -\frac{1.5 \times (-3) \times 3}{1.5(3+3) - 3}$$

$$= \frac{13.5}{6} = 2.25 \text{ cm.}$$

To find the position of the image use the relation.

$$\frac{1}{V} - \frac{1}{U} = \frac{1}{F}$$

Here, $U = u - \alpha = -1.5 - 1.5 = -3.0$

$$\therefore \quad \frac{1}{V} - \frac{1}{3} = \frac{1}{2.25}$$

or $$\frac{1}{V} = \frac{1}{2.25} - \frac{1}{3} = \frac{25}{225} = \frac{1}{9}$$

or $V = 9$ cm

Since $v - \beta = V, v = V + \beta = 9 - 1.5 = 7.5$ cm,

that is, image is on the other side at a distance 75 cm from the nearest surface.

Example : *A plano-convex lens 3 cm thick is made of glass of refractive index 150. If the first surf ace has a radius 3 cm, find (a) the focal length of the lens, (b) the power P and (c) the distances from the first vertex to the two principal planes.*

Here $r_1 = +3$ cm ; $r_2 = \infty$, $t = 3$ cm, $\mu = 1.5$.

$$\therefore \quad \frac{1}{f_1} = -\frac{\mu - 1}{r_1} = -\frac{1.5-1}{+3} = -\frac{0.5}{3} = -\frac{1}{6}.$$

$$\frac{1}{f_2} = -\frac{\mu - 1}{r_2} = -\frac{1.5-1}{\infty} = 0$$

(*a*)
$$F = -\frac{\mu f_1 f_2}{\mu(f_2 - f_1) - t} = -\frac{\mu}{\frac{\mu}{f_1} - \frac{\mu}{f_2} - \frac{t}{f_1 f_2}}$$

$$= -\frac{1.5}{-\frac{1.5}{6}} = +6 \text{ cm.}$$

(*b*) Power $P = 100\ /f = +\ 100/6 = +\ 16.67$ D.

(*c*)
$$\alpha = O_1H_1 = -\frac{f_1 t}{\mu(f_2 - f_1) - t}$$

$$= -\frac{t/f_2}{\left(\frac{\mu}{f_1} - \frac{\mu}{f_2}\right) - \frac{t}{f_1 f_2}} = 0 \quad \ldots\left[\because \frac{1}{f_2} = 0\right]$$

$$\beta = O_2H_2 = -\frac{f_2 t}{\mu(f_2 - f_1) - t}$$

$$= -\frac{t/f_1}{\left(\frac{\mu}{f_1} - \frac{\mu}{f_2}\right) - \frac{t}{f_1 f_2}} = 0$$

$$= -\frac{3\times(1/6)}{-1.5/6} = \frac{+1/2}{-3/12}$$

$$= -\frac{1}{2}\times\frac{12}{3} = -2 \text{ cm}$$

$$O_1H_1 = \alpha = 0;\ O_1H_2 = O_1O_2 - H_2O_2$$

$$= O_1O_2 + O_2H_2 = 3 - 2 = +1 \text{ cm}.$$

That is distances from the first vertex to the first and the second principal points are zero and 1 cm respectively.

For Ready Reference

(a) For a thick lens of thickness t, refractive index m and with f_1 and f_2 as the first focal distance of the first surface and second focal distance of second surface respectively,

$$F = -\frac{\mu f_1 f_2}{\mu(f_2 - f_1) - t}$$

$$\beta = -\frac{f_2 t}{\mu(f_2 - f_1) - t}$$

and $$\alpha = -\frac{f_1 t}{\mu(f_2 - f_1) - t}.$$

(b) For Two thin lenses of focal lengths, f_1 and f_2 separated by a distance t

$$F = +\frac{f_1 f_2}{f_2 + f_1 - t}$$

$$\beta = -\frac{f_2 t}{f_1 + f_2 - t}$$

and $$\alpha = +\frac{f_1 t}{f_1 + f_2 - t}.$$

In calculating F, β and α in either case, the usual sign conventions to the given quantities must be applied.

5

Thin Lenses

Two Thin Lenses Separated by a Finite Distance : In the theory of optical instruments it is often essential to make use of two or more lenses separated by a distance for the removal of certain aberrations in the image formed by the lens system. We shall, therefore, consider the image formation by such complex systems confining our attention only to paraxial rays and approach this subject by studying the behaviour of two thin lenses having a common axis and separated by some finite distance.

It is possible to replace the combination by a single thin lens which, when placed at a suitable fixed point, produces an image of *the same size but not generally in the same position* as that produced by the combination. It is in this *restricted* sense, namely, that it produces the same transverse magnification as does the combination tor all small objects normal to its axis that the lens is said to be an equivalent lens and its focal length is said to be the equivalent focal length.

Equivalent Focal Length of a Convex Lens System : Let L_1, L_2, be two thin convex lenses of focal lengths f_1, f_2 placed on a common axis a distance 'a' apart.

Consider a ray PA incident on L_1, parallel to the axis at a height $O_1A = h_1$, above the axis. It suffers a deviation through an angle δ_1 at the first lens, and proceeds towards its second principal focus F_{sa} meeting the second lens at a height $O_2B = h_2$, where it is further deviated through an angle δ_2. It then intersects the axis in F_2 the second principal focus of the lens system.

Let PA produced, cut F_2B in M_2, then a single convex lens placed in the position M_2H_2 and having focal length H_2F_2 is equivalent to the lens system. Thus, H_2F_2 (= F) is the required equivalent focal length, we have

$$\text{deviation by first lens, } \delta_1 = \frac{h_1}{f_1},$$

$$\text{deviation by second lens, } \delta_2 = \frac{h_2}{f_2},$$

$$\text{anddeviation by equivalent lens, } \delta = \frac{h_1}{F}.$$

Also $\delta = \delta_1 + \delta_2.$

$$\therefore \quad \frac{h_1}{F} = \frac{h_1}{f_1} + \frac{h_2}{f_2}. \quad \dots (i)$$

Now, $h_2 = O_2B = O_2K - BK$

$$= h_1 - a\delta_1 = h_1 - a\frac{h_1}{f_1}$$

$$= h_1\left(1 - \frac{a}{f_1}\right).$$

Putting the value of h_2 in (i), we have

$$\frac{h_1}{F} = \frac{h_1}{f_1} + \frac{h_1}{f_2}\left(1 - \frac{a}{f_1}\right).$$

or $$\frac{1}{F} = \frac{1}{f_1} + \frac{1}{f_2} - \frac{a}{f_1 f_2} \quad \text{... (1)}$$

$$= \frac{f_1 + f_2 - a}{f_1 f_2} \quad \text{... (1a)}$$

whence $$F = +\frac{f_1 f_2}{f_1 + f_2 - a} \quad \text{... (2)}$$

or $$F = \frac{-f_1 f_2}{\Delta},$$

where $\Delta = a - (f_1 + f_2)$ and is known as the Optical separation or optical interval between the two lenses.

Let us find the position of the equivalent lens, *i.e.*, distance O_2H_2 or O_1H_3.

The Δs $BF_2\ O_2$ and $M_2F_2H_2$ are obviously similar.

$$\therefore \quad \frac{h_2}{h_1} = \frac{O_2F_2}{H_2F_2}$$

or $$O_2F_2 = \frac{h_2}{h_1} .$$

$$H_2F_2 = \left(1 - \frac{a}{f_1}\right) F \text{. ...} \left[\because \frac{h_2}{h_1} = 1 - \frac{a}{f_1}\right].$$

Now, $$O_2H_2 = H_2F_2 - O_2F_2$$

$$= F - F\left(1 - \frac{a}{f_1}\right) = F\frac{a}{f_1}$$

$$= \frac{f_1 f_2}{f_1 + f_2 - a} \times \frac{a}{f_1} = \frac{af_2}{f_1 + f_2 - a}.$$

Let us conventionally put it equal to $-\beta$ as O_2H_2 is to the left of lens O_2.

$$\therefore \qquad -\beta = \frac{af_2}{f_1 + f_2 - a},$$

or
$$\beta = -\frac{af_2}{f_1 + f_2 - a} = -\frac{aF}{f_1} \qquad \text{... (3)}$$

and
$$O_1H_2 = O_1O_2 - O_2H_2$$

$$= a - \frac{Fa}{f_1} = a\left(1 - \frac{F}{f_1}\right)$$

Thus, the equivalent lens must be placed at a distance $-\frac{aF}{f_1}$ in front of the second lens or a distance $a\left(1 - \frac{F}{f_1}\right)$ behind the first lens.

The plane M_3H_2 where the deviation of the incident parallel ray coming from the left appears to occur is called the second principal plane of the lens system and the point H_2, where it cuts the axis, the second principal point. *The second principal focus F_2 is situated at a distance P towards the right of this point.*

In the same way by considering a ray parallel to the axis incident from the right-hand side we could find the position of F_1 the point where it intersects the principal axis after emerging from the lens system. It is called the first principal focus of the lens system. In the same way as above the position of M_1H_1, the first principal plane and hence the point H_1, the first principal point of the lens system could be determined.

As above we will find that

$$F = +\frac{f_1 f_2}{f_1 + f_2 - a}$$

$$O_1H_1 = \frac{Fa}{f_2} = \frac{f_1 f_2}{f_1 + f_2 - a} \times \frac{a}{f_2} = \frac{af_1}{f_1 + f_2 - a}$$

Let this distance be denoted by α, then

$$\alpha = \frac{af_1}{f_1 + f_2 - a} = + \frac{af}{f_2} \qquad ...4$$

and $$O_1H_1 = a\left(1 - \frac{F}{f_2}\right).$$

We thus find that the equivalent focal length is independent of the direction from which the light enters the lens system. It also becomes evident that the behaviour of the system cannot be reproduced in all respects by an equivalent thin lens in a fixed position, for the equivalent lens would need to be placed in the plane M_2H_2 to deal with a ray parallel to the axis from the left while it must be placed in the plane M_1H_1 to deal with a ray parallel to the axis from the right in the way as the system does. *The first principal focus* F_1 *is situated at a distance F towards the left of this point* H_1.

(a) *Power of the Coaxial Systems :* If P_1 and P_2 are the powers of the component lenses and P the power of the combination, then by equation (1)

$$P = P_1 + P_2 - aP_1P_2 \qquad ...5$$

Equivalent Focal Length of a Concave Lens System : Let, as before, L_1, L_2, be two thin concave lenses placed on a common axis at a distance *'a'* apart.

Consider a ray *PA* incident on L_1 parallel to the axis at a height $O_1 A = h_1$ above the axis. It suffers a deviation δ_1 at the first lens and on diverging appears to proceed from its second principal focus F_{2a}. Meeting the second lens at a height $O_2B = h_2$, it is further diverged through δ_2 and appears to intersect the axis in F_z, the second principal focus of the lens system.

Let PA produced cut F_2B in M_2, then a single concave lens placed in the position M_2H_2 and having focal length H_2F_2 is equivalent to the lens system. Thus, H_2F_2 (=F) is the required equivalent focal length, we have

$$\text{deviation by first lens, } \delta_1 = \frac{h_1}{f_1}$$

$$\text{deviation by second lens, } \delta_2 = \frac{h_2}{f_2}$$

$$\text{and deviation by equivalent lens, } \delta = -\frac{h_1}{F}.$$

Also $$\delta = \delta_1 + \delta_2,$$

$$\therefore \quad -\frac{h_1}{F} = -\frac{h_1}{f_1} - \frac{h_2}{f_2},$$

or $$\frac{h_1}{F} = \frac{h_1}{f_1} + \frac{h_2}{f_2} \qquad \text{... (1)}$$

Now, $$h_2 = O_2B = O_2K + KB$$

$$= h_1 + a\delta_1$$

$$= h_1 - a\frac{h_1}{f_1} = h_1\left(1 - \frac{a}{f_1}\right).$$

Putting the value of h_2 in (i), we have

$$\frac{h_1}{F} = \frac{h_1}{f_1} + \frac{h_1}{f_2}\left(1 - \frac{a}{f_1}\right)$$

or $$\frac{1}{F} = \frac{1}{f_1} + \frac{1}{f_2} - \frac{a}{f_1 f_2},$$

whence $$F = +\frac{f_1 f_2}{f_1 + f_2 - a}.$$

If P_1 and P_2 be the powers of the component lenses and P the power of the combination, then

$$P = P_1 + P_2 - aP_1 P_2.$$

These relations are the same as expressed in equations (1), (2) and (5) in the case of the convex lens system.

The determination of positions of the principal planes and principal points in this case is left as an exercise for the students.

All the formulae derived above in respect of convex and concave systems also hold true when the component lenses are of opposite sign.

Relation between the positions of conjugate points and equivalent focal length in a coaxial lens system separated by a finite distance. The usual thin lens formula

$$\frac{1}{V} - \frac{1}{U} = \frac{1}{F}.$$

also holds in the case of coaxial lens system consisting of two or more lenses not in contact provided, however, U and V are measured from first and second principal points H_1 and H_2 respectively, *viz.*, U = u – α, and V= v – β, where α is the distance of the first principal plane from the first lens and β is the distance of the second principal plane from the second lens, u and v are the distances of the object from the first lens and that of image from the second lens respectively. The distances U and V are called the reduced object and image distances respectively.

Let f_1 and f_2 be the focal lengths of the two component lenses separated by a distance a. Consider an object u cm away from the first lens. Then, if v' be the distance of the image formed by this lens and v is the distance of the final image from the second lens, we can write (remembering that the image due to the first lens is the object for the second),

$$\frac{1}{v'} - \frac{1}{u} = \frac{1}{f_1}$$

and $$\frac{1}{v} - \frac{1}{v' - a} = \frac{1}{f_2}.$$

Eliminating v' from these equations, we have

$$\frac{1}{v - \beta} - \frac{1}{u - \alpha} = \frac{1}{F}, \quad \text{... (6)}$$

or $$\frac{1}{V} - \frac{1}{U} = \frac{1}{F} \quad \text{... (7)}$$

where $$F = +\frac{f_1 f_2}{f_1 + f_2 - a},$$

$$\beta = -\frac{f_2 a}{f_1 + f_2 - a},$$

$$\alpha = +\frac{f_1 a}{f_1 + f_2 - a}.$$

The above relations can be deduced by algebra as below :

Now, $$\frac{1}{v'} - \frac{1}{u} = \frac{1}{f_1} \quad \text{... (i)}$$

and $$\frac{1}{v} - \frac{1}{v' - a} = \frac{1}{f_2}. \quad \text{... (ii)}$$

Let us eliminate v'

From (*i*) $$\frac{1}{v'} = \frac{1}{f_1} + \frac{1}{u} = \frac{u + f_1}{u f_1},$$

$\therefore$ $$v' = \frac{u f_1}{u + f_1} \quad \text{... (iii)}$$

From (*ii*) $$\frac{1}{v' - a} = \frac{1}{v} - \frac{1}{f_2} = \frac{f_2 - v}{v f_2},$$

or $$v' - a = \frac{v f_2}{f_2 - v},$$

$\therefore$ $$v' = \frac{v f_2}{f_2 - v} + a \quad \text{... (iv)}$$

Equating the values of v' in (*iii*) and (*iv*)

$$\frac{uf_1}{u+f_1} = \frac{v f_2}{f_2 - v} + a .$$

Multiplying throughout by $(u + f_1)(f_2 - v)$, we have

$$uf_1 (f_2 - v) = vf_2 (u + f_1) + a (u + f_1)(f_2 - v)$$

Opening the brackets and collecting the terms in uv, u and v

$$uf_1f_2 - uvf_1 = uvf_2 + vf_1f_2 + uaf_2$$
$$- uva + af_1f_2 - vaf_1,$$

or
$$uvf_1 + uvf_2 - uva + uaf_2 - uf_1f_2 - vf_1f_2$$
$$- vaf_1 + af_1f_2 = 0$$

or $uv\left[f_1 + f_2 - a\right] + u\left[af_2 - f_1f_2\right] + v\left(f_1f_2 - af_1\right) + af_1f_2 = 0$

Diyiding by $(f_1 + f_2 - a)$ throughout, we get

$$uv + \frac{u\left(af_2 - f_1f_2\right)}{f_1 + f_2 - a} + \frac{v\left(f_1f_2 - af_1\right)}{f_1 + f_2 - a} + \frac{af_1f_2}{f_1 + f_2 - a} = 0. \quad \text{... } (v)$$

This equation can be put in the form

$$\frac{1}{v - \beta} - \frac{1}{u - \alpha} = \frac{1}{F}. \quad \text{... } (vi)$$

Multiplying (*vi*) throughout by $F (u - \alpha)(v - \beta)$ and collecting terms in uv, u and v, we obtain

$$uv - (F + \beta)u + (F - \alpha) v - F(\beta - \alpha) + \alpha\beta = 0.... \quad (vii)$$

In order that (*v*) and (*vii*) should be identical, we must have

$$- F - \beta = \frac{af_2 - f_1f_2}{f_1 + f_2 - a}, \quad \text{... } (viii)$$

$$F - \alpha = \frac{f_1 f_2 - a f_1}{f_1 + f_2 - a} \quad \text{...}(ix)$$

and $$-F\beta + F\alpha + \alpha\beta = \frac{a f_1 f_2}{f_1 + f_2 - a}. \quad \text{...}(x)$$

Thus *(viii)*, *(ix)* and *(x)* are simultaneous equations to find the three unknowns α, β and *F*.

To solve these, first multiply the corresponding sides of equations (*viii*), and (*ix*). This gives us

$$-F^2 - F\beta + F\alpha + \alpha\beta = \frac{(af_2 - f_1 f_2)(f_1 f_2 - af_1)}{(f_1 + f_2 - a)^2} \quad \text{... }(xi)$$

Substracting (*xi*) from (*x*) gives

$$F^2 = \frac{a f_1 f_2}{f_1 + f_2 - a} - \frac{(af_2 - f_1 f_2)(f_1 f_2 - af_1)}{(f_1 + f_2 - a)^2}$$

$$= \frac{af_1 f_2 (f_1 + f_2 - a) - (af_2 - f_1 f_2)(f_1 f_2 - af_1)}{(f_1 + f_2 - a)^2}$$

$$= \frac{(f_1 f_2)^2}{(f_1 + f_2 - a)^2}$$

$$\therefore \qquad F = \pm \frac{f_1 f_2}{f_1 + f_2 - a} \quad \text{... }(xii)$$

In order to find whether the positive or negative sign is to be retained with the values of *F*, it must be noted that equation (*xii*) must be true for all values of *a*.

When $a = 0$, from (*i*) and (*ii*), we have

$$\frac{1}{v} - \frac{1}{u} = \frac{1}{f_1} + \frac{1}{f_2} = \frac{1}{F},$$

or $$F = \frac{f_1 f_2}{f_1 + f_2}.$$

Substituting $a = 0$ in (*xii*) we see that we obtain the value of F just found provided we take the positive sign. Hence

$$F = +\frac{f_1 f_2}{f_1 + f_2 - a} \qquad \text{... (xiii)}$$

Adding (*viii*) and (*xiii*), we have

$$-\beta = \frac{af_2 - f_1 f_2}{f_1 + f_2 - a} + \frac{f_1 f_2}{f_1 + f_2 - a}$$

$$= \frac{af_2}{f_1 + f_2 - a},$$

or $$\beta = -\frac{af_2}{f_1 + f_2 - a} = -\frac{aF}{f_1}. \qquad \text{... (xiv)}$$

Substracting (*ix*) from (*xiii*), we get

$$\alpha = \frac{f_1 f_2}{f_1 + f_2 - a} - \frac{f_1 f_2 - af_1}{f_1 + f_2 - a}$$

$$= \frac{af_1}{f_1 + f_2 - a} = \frac{aF}{f_2} \qquad \text{... (xv)}$$

Hence, also $$\alpha f_2 = \beta f_1 \qquad \text{... (xvi)}$$

which is an important relation.

Here α is the distance of first principal plane from first lens and β the distance of second principal plane from second lens.

Transverse Magnification

For a combination of two lenses, the magnification is obtained by the ratio of the reduced image distance from the second principal plane to the reduced object distance from the first principal plane.

i.e.,
$$m = \frac{\text{reduced image distance}}{\text{reduced object distance}} = \frac{V}{U} \quad \text{... (8)}$$

Or, *alternatively*, if u_1 and v_1 are the distances of the object and the intermediate image from the first lens and u_2 and v_2 the distances of the intermediate image and the final image from the second lens, we write

$$\text{for first lens, } m_1 = \frac{v_1}{u_1}; \quad \text{for second lens, } m_2 = \frac{v_2}{u_2}.$$

From which magnification for the combination is given by

$$m = m_1 \times m_2 = \frac{v_1}{u_1} \times \frac{v_2}{u_2}. \quad \text{... (9)}$$

Significance of the Principal Points and Principal Foci : The principal points H_1, and H_2 of a lens system referred to in the preceding sections are a pair of conjugate points for which transverse magnification is plus unity or $m = +1$. This can be easily proved, where two rays, one originally parallel to the axis and the other through F_1, the first principal focus of the system are drawn in such a manner that the height above the axis of the incident parallel ray which passes on emergence through F_2, the second principal focus of the system, is the same as that of the emergent parallel ray which originated at F_1.

It is now obvious that the pair of rays converging to M_1, before entering the system becomes the pair of rays diverging from M_2 after emergence. According to our usual way of locating the image of an object point, the course of two rays from a given object point through the system to their final intersection on emergence, gives rise to the image point.

Therefore, M_1 and M_2 are conjugate points of the system and the transverse planes M_1H_1 and M_2H_2 are conjugate planes. They are evidently planes of unit positive magnification, since the transverse magnification is unity, h_1 being equal to h_2. In other words these planes are such that a small object placed normal to the axis at H_1 will have an image equal to itself and similarly placed at H_2. They are called the unit planes or principal planes of the system. It is the points of intersection H_1 and H_2 of these planes with the principal axis that have been named as unit points (points of unit magnification) or the principal points of the system.

The positions of these points, of course, will depend upon the particular set of lenses. It is from these points that the distances may be measured in all cases, that is, the distance of the object from the first principal point and that of the image from the second principal point in order that the simple thin lens formula may be applied as usual. It is in fact for this reason that these points are called principal points and planes through them normal to the axis of the system principal planes. There is another important property of these points, *viz* , that when a lens system is situated in a uniform medium a ray directed towards the first point will, after refraction through the lens system, appear to proceed from the second point and parallel to the incident ray.

Just as for a single lens so for a lens system there are principal foci and principal focal planes. The first principal focus is at a distance F towards the left of the first principal point H_1, and the second principal focus is at the same distance towards the right of the second principal point H_2 .They are generally denoted by F_1 and F_2 respectively. The principal foci possess the usual properties, namely, that rays directed so as to pass through the first principal focus F_1 at incidence on the first lens will emerge through the second lens parallel to the axis and rays incident on the first lens parallel to the axis will emerge from the second lens in a direction passing through the second principal focus F_2. The planes passing through the principal foci and at right angles to the axis are called principal focal planes.

Numerically, the first focal length H_1F_1 and the second focal length H_2F_2 are equal, each being equal to F. But they are of opposite sign. According to our convention H_2F_2 is *positive* and H_1F_1 is *negative,*

i.e., $$H_2F_2 = -H_1F_1$$

The properties of the principal planes and the focal planes are frequently used in the treatment of eye-pieces, in the complicated lens system such as are employed in the construction of objectives of high power microscopes.

The Principal Points of a Coaxial Lens System are Conjugated Foci : This can also be easily proved by considering the relation

$$\frac{1}{v-\beta}-\frac{1}{u-\alpha}=\frac{1}{F}.$$

By transposing, we have

$$\frac{1}{v-\beta}=\frac{1}{F}+\frac{1}{u-\alpha}=\frac{(u-\alpha)+F}{F(u-\alpha)},$$

or $$v-\beta=\frac{F(u-\alpha)}{(u-\alpha)+F} \quad \text{... (10)}$$

When $u = \alpha$, $v - \beta = 0$ or $v = \beta$ for all values of F. Hence, an object at the first principal point will produce an image at the second principal point.

Nodal Points

Besides the principal foci and principal points, which are of special importance in developing the theory of coaxial systems of thin lenses, there is another important pair of conjugate points on the axis of such an optical system which may be used to simplify the solution of certain problems. They are so situated that an incident ray directed towards the first will, on emerging from the lens system, proceed (or appear to proceed) from the second point in a direction parallel to that of the incident ray. These points are called the first and second nodal points of the system and are denoted by N_1 and N_2 respectively. They are the points for which the angular magnification is + 1, for the incident ray and its conjugate emergent ray make the same angles with the axis. *When the medium on both sides of the* **lens** *system is the same the nodal points and the principal points are coincident.* This can be easily shown by making use of Helmholtz Lagrange's equation according to which

$$\mu_1 h \alpha_1 \equiv \mu_2 h_2 \alpha_2.$$

where the letters have their usual significance. If the medium on both sides is air,

$\mu_1 = \mu_2 = 1$; and since $\alpha_1 = \alpha_2$, we have

$h_1 = h_2$, *i.e.*, the transverse magnification is also unity.

Hence the nodal points for a combination of lenses placed in air coincide with the principal points. The planes transverse to the principal axis through N_1, and N_2 are called nodal planes- Since nodal points are conjugate, the nodal planes are also conjugate.

The Cardinal Points of a Lens System : In the study of the coaxial systems of lenses we have discussed above some points which have an important bearing on the theory of such systems, namely, the two principal foci, the two principal or unit points and the two nodal points. These six points are called the cardinal points or Gaussian points after the name of a German optician E.F. Gauss, who first investigated the behaviour of thick lenses and coaxial systems of lenses He showed how these points have enabled us to treat optical systems as units and to formulate in their case relationships similar to those for thin lenses or single surfaces giving positions and sizes of the images directly in terms of those of the objects used. By their use we can conveniently avoid the tedious process of treating a coaxial system as a succession of spherical surfaces and considering the image formed by the first surface as the object for the following one and so on, ultimately to determine the size and location of the final image formed by the system.

Besides the cardinal points there are two more pairs of conjugate points of lesser importance with negative linear and angular magnifications. They are respectively the *anti-principal* and *anti-nodal* points.

Graphical Construction of the Image of an Object on the Axis of a Coaxial System : To construct the image of an object in a system of two lenses we may proceed to trace the path of rays through each lens separately. Such a method is rather tedious, so in practice the actual positions of the lenses are abandoned and we make use of the properties of the principal foci and principal planes which are planes of unit magnification as explained above, *viz.*, the emergent ray proceeds or appears to proceed from a point in the second principal plane at the same distance from the principal axis of the system and on the same side of it as is the point in the first principal plane to which the incident ray proceeds or appears to proceed. Knowing therefore the positions of the principal foci and unit points, we can easily construct the image of a small object on the axis of a coaxial system.

Let us consider in turn a system of two coaxial lenses in which the focal length is (a) positive, and (*b*) negative.

When the Combination is Positive : Let *PQ* denote a small object at right angles to the principal axis at *Q*, the cardinal points F_1, F_2, H_1 and H_2 of the system being as indicated. Then a ray PA_1, parallel to the axis meeting the first principal plane in A_1, will emerge through a point A_2 in the second principal plane such that $H_2A_2 = H_1 A_1$ and will pass through F_2, the second principal focus of the lens system. Also a ray PB_1 passing through F_1 and meeting the first principal plane in B_1 will emerge from the system in the direction B_2P_1 parallel to the axis so that $H_2B_2 = H_1B_1$. These two rays intersect in P_1, the image of P, and P_1Q_1 drawn perpendicular, to the axis is the image of PQ. The image is real but inverted.

When the combination is negative. Let PQ as before be a small object perpendicular to the axis of the lens system. The paths of two rays from P, one PA_1 parallel to the axis and the other PB_1 directed towards F_1 are traced in the usual way so that the first ray emerges out of the system at a point A_2 and appearing to proceed from F_2 and the second ray on emergence becomes parallel to the axis appearing to intersect the first ray at P_1. P_1Q_1 is the image of PQ. It is virtual, erect and small in size.

It may also be noted that an object ray directed towards H_1 has a corresponding image ray whose direction passes through H_2 and is parallel to that of the object ray.

Images in the case of coaxial lens systems are not generally in the same position as those produced by the equivalent lens. Let us consider in *(a)* the image of an object PQ is formed by a two-lens system of positive focal length by the method explained above and in *(b)* the image of the same object is formed by a single lens of the same focal length as the combination.

Conjugate Relations Referred to the Principal Planes of a Coaxial System : Consider a coaxial system with media of refractive indices μ_1 and μ_2 on its either side. *Let* F_1 and F_2 be the principal foci and H_1 and H_2, the principal points of a convergent coaxial system. Let PQ be an object perpendicular to the principal axis at Q. To obtain the image of PQ let us consider two rays:

(1) A ray PA_1 parallel to the axis meeting the first principal plane in A_1 will emerge through the point A_2 in the second principal plane so that $H_2A_2 = H_1A_1$ and will pass through F_2, the second principal focus of the lens system.

(2) The ray PB_1 passing through F_1 and meeting the first principal plane in B_1 will emerge in the direction B_2P_1 parallel to the axis so that $H_2B_2 = H_1B_1$.

These rays intersect in P_1, the image of P, and P_1Q_1 drawn perpendicular to the axis is the real image of PQ. Let U and V denote respectively the distances of the object and the image from the corresponding principal points. According to our convention the signs to be attached to the various distances are as follows :

$$H_1Q = -U,\ H_2Q_1 = V;\ PQ = n_1,\ P_1Q_1 = -h_2,$$

$$H_1F_1 = -\phi_1,\ H_2F_2 = +\phi_2,$$

where ϕ_1 and ϕ_2 are the first and second principal focal lengths of the lens system.

Now, Δs $F_1H_1B_1$ and PA_1B_1 are similar.

$$\therefore \quad \frac{H_1F_1}{A_1P} = \frac{H_1B_1}{A_1B_1} = \frac{H_1B_1}{H_1A_1 + H_1B_1}$$

$$\text{or} \quad \frac{-\phi_1}{-U} = \frac{-h_2}{h_1 - h_2}. \qquad \dots (a)$$

Also, Δs $H_2A_2F_2$ and $B_2A_2P_1$ are similar.

$$\therefore \quad \frac{H_2F_2}{B_2P_1} = \frac{H_2B_2}{B_2A_2} = \frac{H_2A_2}{H_2A_2 + H_2B_2},$$

$$\text{or} \quad \frac{+\phi_2}{V} = \frac{h_1}{h_1 - h_2}. \qquad \dots (b)$$

Adding (*a*) and (*b*)

$$\frac{-\phi_1}{-U} + \frac{\phi_2}{V} = \frac{h_1}{h_1 - h_2} - \frac{h_2}{h_1 - h_2},$$

or
$$\frac{\phi_1}{U} + \frac{\phi_2}{V} = 1. \qquad \text{... (11)}$$

Dividing (*a*) by (*b*), we get

$$\frac{\phi_1}{U} \times \frac{V}{\phi_2} = -\frac{h_2}{h_1}.$$

$$\therefore \text{ transverse magnification } \left(\frac{h_2}{h_1}\right) = -\frac{V\phi_1}{U\phi_2}. \qquad \text{... (12)}$$

Further, a ray PH_1 directed towards the first principal point H_1 must emerge from the second principal point H_2 along H_2 P_1 because H_1 H_2 and P, P_1 are parts of conjugate points.

Let θ_1 and θ_2 be the angles which the rays PH_1 and H_2P_1 make with the principal axis. Then applying Helmheltz Lagrange's law ($\mu_1 h_1 \tan\theta_1 = \mu_2 h_2 \tan\theta_2$) to the principal planes, we get

$$\mu_1 \tan\theta_1 = \mu_2 \tan\theta_2, \qquad \text{... } (c)$$

since $h_1 = h_2$; the transverse magnification of the planes being h_2/h_1 is +1.

Substituting the values of $\tan\theta_1$ and $\tan\theta_2$ in equation (*c*) taking the conventional signs of *U* and *V* into consideration, we have

$$\mu_1 \frac{h_1}{-U} = \mu_2 \frac{-h_2}{V},$$

whence the lateral magnification

$$m = \frac{h_2}{h_1} = \frac{\mu_1}{\mu_2} \cdot \frac{V}{U} \qquad \text{... (13)}$$

Comparing it with equation (12), we get

$$\frac{\phi_1}{\phi_2} = \frac{\mu_1}{\mu_2}. \quad \text{... (14)}$$

Further, combining relations (11) and (14), we get

$$-\frac{\mu_1}{\mu_2}\cdot\frac{\phi_2}{U}+\frac{\phi_2}{V} = 1,$$

or $$-\frac{\mu_1}{\mu_2}\cdot\frac{1}{U}+\frac{1}{V} = \frac{1}{\phi_2},$$

or $$\frac{\mu_1}{V}-\frac{\mu_1}{U} = \frac{\mu_2}{\phi_2}$$

$$= -\frac{\mu_1}{\phi_1}. \quad \text{... (15)}$$

For the lens system in air $\mu_1 = \mu_2$, so the relations (11), (13) and (14) can be written as

$$\frac{1}{V}-\frac{1}{U} = \frac{1}{\phi_2} = -\frac{1}{\phi_1}, \quad \text{... (16)}$$

$$m = \frac{h_2}{h_1} = \frac{V}{U}, \quad \text{... (17)}$$

and $$\phi_2 = -\phi_1. \quad \text{... (18)}$$

If equations (16), (17) and (18) are compared with the corresponding equations for a thin leas, derived in chapter 3, we find that coaxial lens system can be treated as a single unit like a thin lens provided the object and the image distances are measured from the corresponding principal points of the lens system.

Newton's Formula

Connecting Conjugate Foci for a Coaxial Lens System : This is an important conjugate-distance relation with reference to principal focal planes.

F_1, F_2 are the principal foci and H_1, H_2 the unit points. P_1Q_1 is the image of an object PQ, perpendicular to the axis drawn. Let $PQ = h_1$ $P_1Q_1 = h_2$, $F_1Q = x_1$ and $F_2Q_1 = x_2$ and let ϕ_1 and ϕ_2 be the first and second focal lengths of the system. According to our convention h_1, ϕ_2, x_2 are positive and h_2, ϕ_1, x_1 are negative.

Now, Δs PQF_1 and $B_1H_1F_1$ are similar.

$$\therefore \qquad \frac{H_1B_1}{PQ} = \frac{H_1F_1}{F_1Q},$$

$$\text{or} \qquad \frac{-h_2}{h_1} = \frac{-\phi_1}{-x_1} = \frac{\phi_1}{x_1}, \qquad \ldots (i)$$

Also, Δs $A_2H_2F_2$ and $P_1Q_1F_2$ are similar.

$$\therefore \qquad \frac{P_1Q_1}{H_2A_2} = \frac{F_2Q_1}{H_2F_2},$$

$$\text{or} \qquad \frac{h_2}{h_1} = \frac{x_2}{\phi_2}. \qquad \ldots (ii)$$

Comparing equations (*i*) and (*ii*)

$$\frac{\phi_1}{x_1} = \frac{x_2}{\phi_2},$$

$$\text{or} \qquad x_1x_2 = \phi_1\phi_2.$$

When the medium is the same on both sides

$$\phi_1 = -\phi_2,$$

$$\therefore \qquad x_1x_2 = \phi_2^2 = -F^2 \qquad \ldots (19)$$

since F is being used for f_2 in the text.

Also from (*i*) and (*ii*)

$$m = \frac{h_2}{h_1} = \frac{+F}{x_1} = \frac{-x}{F} \qquad \ldots (20)$$

Example : *Two convex lenses with focal lengths 10 cm and 20 cm ore placed co-axially 5 cm apart. Find (a) the equivalent focal length of combination, (b) the positions of the principal planes, focal* *planes, and (c) the position of the image of an object placed 60 cm from the first lens.*

(*a*) Here $f_1 = +10$ cm, $f_2 = +20$ cm and $a = 5$ cm.

$$\therefore \quad H_2F_2 = F = +\frac{f_1 f_2}{f_1 + f_2 - a}$$

$$= +\frac{10 \times 20}{10 + 20 - 5} = +\frac{200}{25} = +8 \text{ cm.}$$

and $\quad H_1F_1 = -8$ cm.

$$O_2H_2 = \beta = -\frac{af_2}{f_1 + f_2 - a}$$

$$= -\frac{5 \times 20}{10 + 20 - 5} = -\frac{100}{25} = -4 \text{ cm.}$$

$$O_1H_1 = \alpha = +\frac{af_1}{f_1 + f_2 - a}$$

$$= +\frac{5 \times 10}{10 + 20 - 5} = +\frac{50}{25} = +2 \text{ cm.}$$

(*b*) Using these values the positions of the principal points and focal points have been indicated

(*c*) Since $u = -60$ cm, we can now write

$$U = u - \alpha = -60 - 2 = -62 \text{ cm,}$$

$$F = +8 \text{ cm}, \; V = ?$$

Applying the relation,

$$\frac{1}{V} - \frac{1}{U} = \frac{1}{F}, \text{ we have } \frac{1}{V} - \frac{1}{-62} = \frac{1}{8}$$

$$\therefore \quad \frac{1}{V} = \frac{1}{8} - \frac{1}{62},$$

whence $$V = \frac{8 \times 62}{54} = 9.2 \text{ cm.}$$

$$v - \beta = 9.2 \text{ cm}$$

or $$v - (-4) = 9.2 \text{ cm.}$$

or $$v = 5.2 \text{ cm.}$$

That is, the final image is 5.2 cm beyond the second lens.

$$m = \frac{V}{U} = \frac{9.2}{-62} = -0.15.$$

Alternatively : The same result may be arrived at from *first* principles, as below :

For the first lens

$$\frac{1}{v_1} - \frac{1}{u_1} = \frac{1}{f_1},$$

where $u_1 = -60$ cm; $f_1 = +10$ cm.

$$\therefore \quad \frac{1}{v_1} - \frac{1}{-60} = \frac{1}{10}$$

or $$\frac{1}{v_1} = \frac{1}{10} - \frac{1}{60} = \frac{6-1}{60} = +\frac{5}{60}$$

or $$v_1 = +12 \text{ cm.}$$

This image serves as a virtual object for the second lens at a distance $(+12 - 5) = +7$ cm from it.

For refraction through the second lens

$$u_2 = +7 \text{ cm}; f_2 = +20 \text{ cm.}$$

Now, $$\frac{1}{v_2} - \frac{1}{u_2} = \frac{1}{f_2},$$

$$\therefore \quad \frac{1}{v_2} = \frac{1}{v_2} - \frac{1}{+7} = \frac{1}{20},$$

or $$\frac{1}{v_2} = \frac{1}{20} + \frac{1}{7} = \frac{7+20}{20 \times 7} = \frac{27}{20 \times 7}$$

$\therefore$ $$v_2 = \frac{20 \times 7}{27} = 5.2 \text{ cm}$$

whence $$m = \frac{v_1}{u_1} \times \frac{v_2}{u_2} = \frac{12}{-60} \times \frac{5.2}{7} = -0.15.$$

The student should carefully note, that the second principal plane is nearer the first lens than the first principal plane. This is of frequent occurrence and should cause no confusion in the mind of the student, because the object distance is always measured from the first principal plane, wherever it is situated.

Example : *An object is placed at a distance of 50 cm in front of a combination of a thin convex lens of focal length 25 cm and a thin concave lens of focal length 15 cm placed 15 cm apart along the common axis. Find (a) the focal length of the combination, (b) the positions of the cardinal points and (c) the transverse magnification.*

(*a*) Here $f_1 = +25$ cm, $f_2 = -15$ cm and $a = 15$ cm.

$\therefore$ $$H_2F_2 = F = +\frac{f_1 f_2}{f_1 + f_2 - a} = +\frac{25 \times (-15)}{25 - 15 - 15}$$

$$= \frac{375}{-5} = +75 \text{ cm.}$$

and $$H_1F_1 = -75 \text{ cm}$$

$$O_2H_2 = \beta = -\frac{af_2}{f_1 + f_2 - a}$$

$$= \frac{15 \times (-15)}{25 - 15 - 15} = -45 \text{ cm.}$$

$$O_1H_1 = \alpha = +\frac{af_2}{f_1 + f_2 + a}$$

$$= + \frac{15 \times 25}{25 - 15 - 15} = - 75 \text{ cm.}$$

(*b*) The positions of the cardinal points are marked. Since the lens system is in air, N_1 and N_2 coincide with H_1 and H_2 respectively.

(*c*) Applying the thin lens equation to the convex lens, we get

$$\frac{1}{v_1} - \frac{1}{-50} = \frac{1}{25}$$

or
$$\frac{1}{v_1} = \frac{1}{25} - \frac{1}{50}$$

whence
$$v_1 = 50.$$

This image serves as the virtual object for refraction at the concave lens. The distance of the image from the concave lens is equal to, $u_2 = + 50 - 15 = + 35$ cm.

Hence for this lens

$$\frac{1}{v_2} - \frac{1}{35} = -\frac{1}{15}$$

or
$$\frac{1}{v_2} = -\frac{1}{15} + \frac{1}{35},$$

whence
$$v_2 = -\frac{105}{4} \text{ cm.}$$

Resultant transverse magnification

$$m = m_1 \times m_2 = \frac{v_1}{u_1} \times \frac{v_2}{u_2}$$

$$= \frac{50}{-50} \times \frac{-105}{4} \times \frac{1}{35} = +\frac{3}{4}.$$

Example : *Two convex lenses, each, of 20 cm focal length are placed 5 cm apart. A tower 100 metres high and 200 metres distant is viewed through them. Find* (a) *the focal length of the equivalent lens, and (b) the position and size of image.*

(*a*) The focal length of the equivalent lens is given by

$$F = +\frac{f_1 f_2}{f_1 + f_2 - a} = +\frac{20 \times 20}{20 + 20 - 5}$$

$$= +\frac{400}{35} = +11.43 \text{ cm.}$$

(*b*) To find the position and size of image let us take one lens at a time.

For first lens, $u_1 = -$ 200 metres $= -$ 20,000 cm and $f_1 = +$ 20 cm.

Substituting the values in the relation,

$$\frac{1}{v_1} - \frac{1}{u_1} = \frac{1}{f_1},$$

we have $$\frac{1}{v_1} + \frac{1}{20000} = \frac{1}{20},$$

or $$\frac{1}{v_1} = \frac{1}{20} - \frac{1}{20000} = \frac{1000 - 1}{20000}$$

$$= +\frac{999}{20000},$$

whence $$v_1 = +\frac{20000}{999} = +\ 20.02 \text{ cm approx.}$$

If u_2 be the distance from the second lens of the image formed by the first,

then $$u_2 = +\ 20.02 - 5.0 = 15.02 \text{ cm.}$$

This serves as an object for the second lens. Substituting this value in the relation

$$\frac{1}{v_2} - \frac{1}{u_2} = \frac{1}{f_2}$$

we get, $\frac{1}{v_2} - \frac{1}{+15.02} = \frac{1}{20}$,

or $$\frac{1}{v_2} = \frac{1}{20} + \frac{1}{15.02}$$

$$= +\frac{15.02 + 20}{20 \times 15.02} = \frac{35.02}{300.4},$$

whence $$v_2 = \frac{300.4}{35.02} = \mathbf{+\ 8.58\ cm}$$

Now, $$m = \frac{v_1}{u_1} \times \frac{v_2}{u_2} = \frac{20.02}{20000} \times \frac{8.58}{15.02} = \frac{171.77}{300400},$$

i.e., $$\frac{\text{size of image}}{\text{size of object}} = \frac{171.77}{300400}$$

$$\therefore \quad \text{size of image} = \frac{171.77}{300400} \times \text{size of object}$$

$$= \frac{171.77}{300400} \times 10000 = 5.71 \text{ cm}.$$

Note : The reader will come across cases of coaxial systems in which it is not necessary that if the component lenses are converging, the system as a whole needs be converging, it may be both converging or diverging depending upon the positions of the lenses with respect to the incident rays. Thus, the terms 'converging' and 'diverging' used in connection with such systems become *misleading* . Let us illustrate this point by two examples.

Example : *Two thin converging lenses of focal lengths 15 cm and 5 cm are placed 10 cm apart each lens facing the incident ray alternately. Find the positions of the cardinal points and trace the course of a parallel ray through them.*

Taking $f_1 = 15$ cm, $f_2 = 5$ cm, and $a = 10$ cm, we have

$$H_2F_2 = F = \frac{f_1 \times f_2}{f_1 + f_2 - a} = \frac{15 \times 5}{15 + 5 - 10} = 7.5 \text{ cm}$$

and $\quad H_1F_2 = -7.5$ cm

$$O_1H_1 = \alpha = \frac{af_1}{f_1 + f_2 - a} = \frac{10 \times 5}{15 + 5 - 10} = 15 \text{ cm}$$

$$O_2H_2 = \beta = \frac{af_2}{f_1 + f_2 - a} = -\frac{10 \times 5}{15 + 5 - 10} = -5 \text{ cm.}$$

Thus, if the lenses are arranged as with the cardinal points marked as shown the nodal points coinciding with the principal points let us trace the path of a ray PA_1 parallel to the axis through the system. If PA_1 produced, cuts the first unit plane in B_1, it will proceed along B_2F_2 after emergence from the system where B_2 is on the second principal plane and $H_2B_2 = H_1B_1$ If B_2F_2 cuts the second lens in A_2, the path of the rays between the lenses is A_1A_2 and the emergent ray is the prolongation of B_1A_2. The emergent ray actually crosses the principal axis outside the system which is, therefore, a 'converging' system.

If, now the system is reversed, *i.e.*, the component lenses interchange their positions, the focal points and the principal points occupy the same positions as before relative to the two leases, the words 'first' and 'second' must be interchanged, for with the given data $f_1 = 5$ cm and $f_2 = 15$ cm, $O_1H_1 = \alpha = 5$ cm and $O_2H_2 = \beta = -15$ cm, F remains unaltered. These new positions of the cardinal points are marked In this case also, proceeding as usual we find the course of the emergent ray, and conclude that although the focal length of the system is still positive, the parallel incident ray on emergence does not cross it at a real point, so that the system is a diverging' one.

***Example** : An eye-piece is formed by two thin convex lenses of focal lengths 3 cm and 8 cm, and a distance 6 cm apart. Calculate the focal length of the eye-piece to indicate the position of the cardinal points of the eye piece with respect to the lenses.*

We have $f_1 = +\ 3$ cm; $f_2 = +\ 8$ cm and $a = 6$ cm.

$$H_2F_2 = F = +\frac{f_1 f_2}{f_1 + f_2 - a}$$

$$= +\frac{(+3)\times(+8)}{3+8-6}$$

$$= \frac{24}{5} = +\ 4.8 \text{ cm.}$$

and $$H_1F_1 = -4\frac{4}{5} \text{ cm.}$$

$$O_1H_1 = \alpha = +\frac{af_1}{f_1 + f_2 - a}$$

$$= +\frac{6\times(+3)}{3+8-6}$$

$$= +\frac{18}{5} = +\ 3.6 \text{ cm.}$$

$$O_2H_2 = \beta = -\frac{af_2}{f_1 + f_2 - a}$$

$$= -\frac{6\times(+8)}{3+8-6}$$

$$= -\frac{48}{5} = -\ 9.6 \text{ cm.}$$

Using these values the positions of principal foci and principal points have been marked. As the lens system is in air the nodal points N_1 and N_2 coincide with the principal points H_1 and H_2 respectively. Hence the positions of all the six points are marked.

If a ray strikes the first lens at A_1 and the first unit plane in B_1, it must emerge in the direction B_2F_2, where B_2 is on the second unit plane such that $H_2B_2 = H_1B_1$. If B_2F_2 cuts the second lens in A_2, the path of the ray between the lenses is $A_1 A_2$ and the emergent ray is the prolongation of $B_2 A_2$.

Here, F_1 is a real focus and F_2 a virtual focus and the combination is divergent for parallel rays falling *first* on the lens of focal length 3 cm. It can similarly be shown that F_1 is a virtual focus and F_2 a real one and the combination is convergent for rays first falling on the lens of focal length 8 cm.

Minimum Distance between the Object and its Real Image for a Positive Focal Length System of two Lenses : If H_1 and H_2 are the principal points of the system and U and V the object and image distances, as measured from H_1 and H_2 respectively, then x the numerical value of the distance between object and image is given by

$$x = U + V + a, \text{ where } a = H_1H_2.$$

Now, for a positive lens, correcting for signs, we have

$$\frac{1}{V} + \frac{1}{U} = \frac{1}{F},$$

$$\therefore \qquad V = \frac{UF}{U - F}.$$

Hence $$x = U + \frac{UF}{U - F} + a \qquad \text{... } (i)$$

For x to be minimum or maximum

$$\frac{dx}{dU} = 1 + \frac{F(U - F) - UF}{(U - F)^2} + 0 = 0.$$

Solving it, we get $U = 2F$ or $U = 0$.

Differentiating it a second time we will find that $U = 2F$, corresponds to a minimum value of x. The solution $U = 0$ is meaningless as it implies, the contact of the object with the first principal plane. Thus putting the first value of U in (*i*), we have

$$x_{\min} = 2F + \frac{2F^2}{F} + a = 4F + a. \quad \text{... (21)}$$

Example : *A thin converging lens of focal length 15 cm and a thin diverging lens of focal length –10 cm are arranged co-axially at a distance 10 cm apart. Determine the positions of cardinal points.*

Here, $f_1 = 15$ cm, $f_2 = -10$ cm, and $a = 10$ cm

$$H_2F_2 = F = \frac{f_1 f_2}{f_1 + f_2 - a} = \frac{15 \times (-10)}{15 - 10 - 10} = +30$$

and $H_1F_1 = -30$ cm.

$$O_1H_1 = \alpha = \frac{af_1}{f_1 + f_2 - a} = \frac{10 \times 15}{15 - 10 - 10} = -30 \text{ cm.}$$

$$O_2H_2 = \beta = -\frac{af_2}{f_1 + f_2 - a} = \frac{10 \times (-10)}{15 - 10 - 10} = -20 \text{ cm.}$$

The positions of the cardinal points. Since the lens system is in air, the nodal points N_1 and N_2 coincide with the principal points H_1 and H_2 respectively.

Experimental Determination of the Cardinal Points of a Positive Coaxial System in Air : *(a) When the principal foci are real and outside the system.* To locate the principal foci of the lens system O_1 O_2, a plane mirror M and a pin PQ to serve as an object [an illuminated cross wires and a screen may be used instead] are suitably mounted on the optical bench. The lens system is shifted with respect to the pin until there is no parallax between the pin and its inverted image. Thus, the position of $F_{1,}$ the first principal focus of the component lens system is found.

The course of rays for the formation of the inverted image P_1Q_1 of the pin PQ placed in the first focal plane of the lens system. The two will be coincident if the plane of the mirror M is normal to the axis. The interested reader should try to understand the path of the rays through the lens system keeping in view the properties of the cardinal points. A ray PA_1 meeting the first principal plane in B_1 becomes parallel to the axis, cuts the second principal plane in B_t and then proceeds parallel to PH_1, intersects the mirror M in E, where it is reflected along EB_2', so that the angle of incidence is equal to the angle of reflection At B_2', a point in the second principal plane, the ray becomes parallel to the axis meeting the first principal plane in $B_,'$. Join B_1' to $P_{1,}$ a point where the line H_1P_1 drawn parallel to EB_2' cuts PQ produced. Then P_1 is the image of the point P. It can also be proved that the final imageP_1Q_1 is *numerically* equal in size to the object PQ, since the Δs PH_1Q and $P_1H_1Q_1$ can easily shown to be congruent.

The position of the second principal focus F_2 is determined in a similar manner by interchanging the positions of the pin and the mirror M. The focal length of the system is next found by any one of the methods given below and the positions of the principal points H_1 and H_2 located, as their distances from the corresponding principal foci are each equal to the focal length. If the system is in air, the positions of the principal points and the nodal points will coincide.

Newton Method of Finding Focal Length : Keeping the system fixed in position and removing the plane mirror, two pins [or an illuminated object and a screen] are arranged one on each side beyond F_1 and F_2 and parallax is removed between one pin and the inverted image of the other. If x_1 and x_2 are the distances of the object and the image respectively from the corresponding principal foci then $x_1 x_2 = -F^2$ whence F may be calculated. The experiment may be repeated for different conjugate points and mean focal length obtained.

Magnification Method : If U and V are the distances of the object and the image from the first and the second principal points respectively, then

$$m = \frac{V}{U} \text{ and } \frac{1}{V} - \frac{1}{U} = \frac{1}{F}.$$

$$\therefore \qquad 1 - m = \frac{V}{F}. \qquad \text{... } (i)$$

To find the magnification a suitable laboratory method may be employed. If by keeping the lens system fixed we arrange the object and the image scales to give magnifications – 1, –2, –3, etc., then from equation (*i*)

$$1 - (-1) = \frac{V_1}{F} = 2,$$

$$1 - (-2) = \frac{V_2}{F} = 3,$$

$$1 - (-3) = \frac{V_3}{F} = 4,$$

$$\therefore \qquad V_2 - V_1 = V_3 - V_2 = ... = F. \qquad \text{... } (ii)$$

Obviously, therefore, the focal length of the lens system is *numerically* equal to the shift of the image (glass) scale when the magnification is increased by unity. The method is, therefore, applicable when the positions of H_1 and H_2 are unknown and hence U or V cannot be measured directly.

(b) When one of the principal foci is within the system. O_1 and O_2, are the component lenses of the system with F_1 lying in between them and M a plane mirror placed behind them normal to the principal axis. L is an additional lens which

produces a real-image of a bright object P on the axis so as to be received by the system which is adjusted until a clear image is formed on the screen at the site of P. The course of rays in this auto-collimating system indicates that the real image formed by L in the absence of the lens system must occupy the same position as F_1 when the system is in position and arranged as above.

To locate F_2 we proceed as in the first case. Next the focal length of the system is found by the magnification method as Newton's method is not applicable, F_1 being inside, and thus the principal points are located. Since the system is in air, the positions of the principal points and the nodal points are the same.

Nodel Slide Method

Nodal Slide Method of Determining Cardinal Points of a Thick Lens or a Lens System : The method is based on the special property of the nodal points, *viz., that a ray incident upon the lens system and proceeding towards the first nodal point* N_1, *gives rise to an emergent ray proceeding from the second nodal point* N_2 *and parallel to the incident ray.* It leads to the following result that "if a parallel beam of light is incident on a convergent lens system forming an image on a screen at its second principal focus the image does not shift laterally when the system is rotated about a vertical axis passing through its second nodal point."

The principle of the method for a positive lens or a lens system is illustrated where a parallel beam is incident on the lens or a lens system. (*a*) The ray 0 along the axis passes through N_1 and N_2 *to* the focus at Q_2. (*b*) The lens has been rotated through a small angle about an axis through N_2 and the same beam of rays passes through the system coming to a focus at the same point Q_2. The ray 1 is now directed towards N_1 and the ray O towards N_2. When projected across from the plane of N_1, to that of N_2 the rays still converge towards Q_2 even though F_2 is now shifted towards one side. Note that ray 1 approaches N, in exactly the same direction that it leaves N_2 according to the property stated above.

The special type of device designed for the purpose of making use of this property for finding the nodal points is called a nodal slide.

Nodal Slide : It consists of a heavy metal upright carrying a table which is capable of rotation about the vertical axis of the upright and the angle of rotation can be recorded, if so desired, on its graduated circular box. The lens or lens system is mounted in the holders provided on the table and may slide along a bar graduated in mm on which the position of the lens system is read. This movement is in addition to the fact that the nodal upright can itself side along the optical bench on which it is mounted. The graduated bar can be shifted parallel to itself and clamped in any desired position.

In carrying out the experiment the position of the lens or lens system is adjusted so that the optical axis of the system is made to intersect the axis of rotation of the table. There is a white screen having a narrow cross-slit S on one side of the nodal slide and a plane mirror M on the other side. Monochromatic light from a source is sent through the slit S adjusted to lie at the secondary focal point of the lens system. On emergence as a parallel beam this light is reflected back on itself by the plane mirror M, fixed normally to the optical axis and passes again through the lens system and is brought to focus at a point on the screen adjacent to S. The nodal slide is now rotated about its axis back and forth and the lens system repeatedly moved along the graduated bar until the rotation produces no motion of the image of the slit; the slit screen must be adjusted for each new position of the lens system so that the image is sharply focused upon it. When this condition is reached, the point N_2 where the axis of rotation intersects the principal axis locates one nodal point. The table is next turned through 180° and the process repeated till the image is again focused on the screen which must not be moved and its position remains unaffected by a small rotation of the table. Thus the position of the second nodal point N_1 is found. When the experiment is performed in air the positions of the nodal points are those of the principal points as well and the distance N_2S is the correct focal length.

By this arrangement, therefore, we can locate the nodal points and hence also the principal points. We can also find the focal length accurately and hence the focal points are also located, that is, all the cardinal points are thus known. We can also study by this arrangement the variation in the positions of the nodal points for different separations of the two lenses of a coaxial system.

6

The Refraction

A lens is a portion of a transparent medium bounded by two spherical surfaces or one plane and the other spherical. The line which passes through the centres of curvature of the two faces or which passes through the centre of curvature of one face and is perpendicular to the other face is called the principal axis of the lens.

The points at which the axis cuts the faces of the lens are the vertices and the distance between them is the thickness of the lens. In this chapter we shall deal with this lenses, *i.e., lenses* of which thickness can be neglected in comparison with object and image distances. Although there are various types of lenses, yet they can be classed into two main groups according as their thickness at the centre is greater or less than at the edges. The former are called converging lenses and the latter diverging lenses.

In considering refraction through a lens we shall confine our attention to rays which make small angles with the axis and adopt the approximations and conventions of sign of the previous chapters.

Refraction through a Thin Lens : Consider a thin convex lens L of refractive index μ placed in air. Let an object P, lie on the principal axis to the left of the lens. If Q' be the position of the image formed by refraction at the first surface, then Q' serves as the virtual object for refraction at the second surface giving rise to the final image at Q.

If O, O' are the vertices, for a thin lens, its thickness $(t) = OO'$ is neglected. Let $OP = u$, $OQ' = v'$ and $O'Q = v$ and r_1, r_2 the radii of curvature of the first and second surface respectively. Then, assuming that a lens of refractive index μ is placed in air. Then for refraction at the first surface, we have

$$\frac{\mu}{v'} - \frac{1}{u} = \frac{\mu - 1}{r_1} \qquad \ldots (a)$$

and for refraction at the second surface measuring the distance of image from O' and keeping in view that rays are now going from a denser to a rarer medium,

$$\frac{\frac{1}{\mu}}{v} - \frac{1}{v' - t} = \frac{\frac{1}{\mu} - 1}{r_2}.$$

Neglecting t, since the lens is thin and simplifying, we have

$$\frac{1}{v} - \frac{\mu}{v'} = \frac{1 - \mu}{r_2} \qquad \ldots (b)$$

Adding (a) and (b), we get

$$\frac{1}{v} - \frac{1}{u} = (\mu - 1)\left(\frac{1}{r_1} - \frac{1}{r_2}\right) \qquad \ldots (1)$$

The points P and Q thus connected, are known as conjugate foci and equation (1) is known as thin lens formula. *Or more generally*, if refractive index of the lens is μ_2 that of the medium on the left and right, μ_1 and μ_2 respectively we have, for refraction at the first and the second surfaces respectively

$$\frac{\mu_2}{v'} - \frac{\mu_1}{u} = \frac{\mu_2 - \mu_1}{r_1} \quad \text{... } (c)$$

and $$\frac{\mu_2}{v} - \frac{\mu_2}{v'} = \frac{\mu_2 - \mu_2}{r_2} \quad \text{... } (d)$$

Adding (*c*) and (*d*), we get

$$\frac{\mu_2}{v} - \frac{\mu_1}{u} = \frac{\mu_2 - \mu_1}{r_1} + \frac{\mu_2 - \mu_2}{r_2} \quad \text{... } (1a)$$

Assuming the lense to be placed in a uniform medium, so that

$$\mu_1 = \mu_2,$$

then $$\frac{\mu_1}{v} - \frac{\mu_1}{u} = \frac{\mu_2 - \mu_1}{r_1} + \frac{\mu_1 - \mu_2}{r_2}$$

$$= (\mu_2 - \mu_1)\left(\frac{1}{r_1} - \frac{1}{r_2}\right).$$

.or $$\frac{1}{v} - \frac{1}{u} = \left(\frac{\mu_2 - \mu_1}{\mu_1}\right)\left(\frac{1}{r_1} - \frac{1}{r_2}\right). \quad \text{... } (1b)$$

The reader is expected to derive the relations 1, (1*a*) and (1*b*) for a concave lense himself.

First and Second Principal Foci of a Thin Lens : The limiting positions of the object and the image formed by refraction an infinite distance are called the an infinite distance are called the first and second principal foci of lens. These points are denoted by F_1 and F_2 and their distances from the centre of the lens are known as the first and the second local lengths of the lens. The first and the second principal foci of a convex and a concave lens respectively. The foci are *real* for a convex lens and *virtual* for a concave lens. Denoting their distances from the centre of the lens by f_1 and f_2, we have

$u = f_1$, when $v = \infty$ and $v = f_2$, when $u = \infty$.

Putting these values in equation (1), we have

$$\frac{1}{\infty}-\frac{1}{f_1} = (\mu-1)\left(\frac{1}{r_1}-\frac{1}{r_2}\right)$$

or
$$-\frac{1}{f_1} = (\mu-1)\left(\frac{1}{r_1}-\frac{1}{r_2}\right) \quad \text{... (2)}$$

and
$$\frac{1}{f_2}-\frac{1}{\infty} = (\mu-1)\left(\frac{1}{r_1}-\frac{1}{r_2}\right)$$

or
$$\frac{1}{f_2} = (\mu-1)\left(\frac{1}{r_1}-\frac{1}{r_2}\right) \quad \text{... (2a)}$$

From these equations it is obvious that the first and second focal lengths of a lens are *numerically* equal when the lens is placed in a uniform medium. In the subsequent sections, when we speak of the focal length of a lens, we shall mean the second local length when the lens is placed in air and denote it by f. Hence, equation (1) may be written as

$$\frac{1}{v}-\frac{1}{u} = \frac{1}{f} = (\mu-1)\left(\frac{1}{r_1}-\frac{1}{r_2}\right) \quad \text{... (3)}$$

Equation (3) gives a fundamental thin lens relation expressing the focal length in terms of the radii of curvature and the refractive index of the material of the lens and is often known as Lens makers' formula.

Since $\mu > 1$, the sign of focal length depends upon that of

$$\left(\frac{1}{r_1}-\frac{1}{r_2}\right).$$

If a lens of refractive index μ_2 is placed in a uniform medium of refractive index μ_1, the focal length is given by

$$\frac{1}{f} = \left(\frac{\mu_2-\mu_1}{\mu_1}\right)\left(\frac{1}{r_1}-\frac{1}{r_2}\right) \quad \text{... (3a)}$$

and as a consequence of equations (2) and (2*a*) for the first and the second principal focal lengths respectively, the above equation may be written as

$$\frac{f_2}{v}+\frac{f_1}{u} = 1. \qquad \text{... (4)}$$

If, however, a lens of refractive index μ_2 is placed in a medium of refractive index μ_1 on the left and that of refractive index μ_3 on the right, then keeping in mind the definitions of the first and second focal lengths we can easily deduce from equation (1*a*) that

$$\frac{\mu_3}{v}-\frac{\mu_1}{u} = -\frac{\mu_1}{f_1} = \frac{\mu_3}{f_2} \qquad \text{... (5)}$$

and

$$\frac{f_1}{f_2} = -\frac{\mu_1}{\mu_2} \qquad \text{... (5a)}$$

That is, the first *and the second focal lengths are in the ratio of the refractive indices of the media in which the coi responding focal points lie.*

Power of a Lens

The expression

$$(\mu-1)\left(\frac{1}{r_1}-\frac{1}{r_2}\right),$$

which by equation (3) is equal to the reciprocal of the focal length, is called the power of the lens. Denoting it by P, we have

$$P = \frac{\mu-1}{r_1}+\frac{1-\mu}{r_2}, \qquad \text{... (6)}$$

from which it follows that the *power of a lens is the sum of the powers of its refracting snrfaces.*

According to the conventions of sign used in the text, for a bi-convex lens in air, r_1 is positive and r_2 is negative, whence the focal length and power of a convex lens are positive. For a biconcave lens in air, r_1 is negative and r_2 is positive, so the focal length and power are both negative.

In practical optics a lens is of *unit* power if it brings parallel rays to focus at a distance of 1 metre. This unit of converging or diverging power is called the dioptre. Thus a converging lens of 1 metre focal length is said to have a power +1 *D* and that of 25 cm focal length + 4D. Similarly, a concave lens of 25 cm focal length has a power – 4*D* and so on.

Optical Centre of a Lens : Let C_1 and C_2 be the centres of curvature of the faces of a bi-convex lens placed in a uniform medium so that the principal axis cuts the leas surfaces in O_1 and O_2. Draw two parallel radii C_1Q and C, R intersecting the leas surfaces in *Q* and *R* respectively, then the tangents at *Q* and *R* will obviously *be* parallel to each other. Hence, a ray *PQ* which on refraction traverses the path *QR* will emerge parallel to *PQ*. The point *O*, where the ray *QR* cuts the axis, is called the optical centre of the lens. To determine the position of *O* consider the similar Δ*s* C_1PQ and C_2OR

Here, $$\frac{OC_1}{OC_2} = \frac{C_1Q}{C_2R} = \frac{C_1C_2}{O_2C_2} = \frac{r_1}{r_2}.$$

Also since $$\frac{O_1C_1}{O_2C_2} = \frac{OC_1}{OC_2} = \frac{O_1C_1 - OC_1}{O_2C_2 - OC_2} = \frac{O_10}{O_20}.$$

$$\therefore \quad \frac{O_10}{O_20} = \frac{r_1}{r_2}. \qquad \text{... (7)}$$

that is, for a given lens the position of *O* is fixed. Hence, any ray which emerges from the lens in a direction parallel to the incident ray, passes through the point *O* or *vice versa*. If the lens is thin, we can almost say that a ray passing through the optical centre goes straight. This property is used in the construction of the images of objects.

Construction of the Image of a Small Object by a Lens: Let PQ, be a small object placed perpendicular to the axis of a lens. A ray PA incident parallel to the axis, after refraction proceeds in the direction AF_2 or F_2A and a ray PB incident in the direction PF_1, after refraction becomes parallel to the axis. Let these two rays intersect or appear to do so, then P' is the image of P and $P'Q'$ perpendicular to the axis is the image of PQ. Since a ray passing through the optical centre O of the lens suffers no deviation, P and P' will also lie on the line passing through

(a) *Transverse or Lateral Magnification :* It is defined as the ratio of any transverse dimension of the image to the corresponding transverse dimension of the object. Denoting it by m, we have

$$m = \frac{h_2}{h_1} = -\frac{P'Q'}{PQ} = -\frac{BO}{PQ} = -\frac{F_1O}{F_1Q}$$

$$= -\frac{F_1O}{OQ - OF_1} = -\frac{+f}{-u-f} = \frac{f}{u+f} \quad \dots (8)$$

Also,
$$\mu = \frac{P'Q'}{AO} = -\frac{F_2Q'}{F_2O} = -\frac{OQ' - OF_2}{F_2O}$$

$$= \frac{v-f}{f} = \frac{f-v}{f} \quad \dots (9)$$

Thus, transverse magnification varies *inversely* as the distance of the object or *directly* as the distance of the image from the corresponding focus.

Another expression for the transverse magnification is

$$m = \frac{h_2}{h_1} = \frac{P'Q'}{PQ} = -\frac{OQ'}{OQ} = \frac{v}{-u} = \frac{v}{u} \quad \dots (10)$$

This result is also obtained by considering the magnification at each surface. If m_1 and m_2 be their respective magnifications, then

$$m_1 = \frac{v'}{u} \div \frac{u}{l} = \frac{v'}{\mu u}$$

and $$m_2 = \frac{v}{l} \div \frac{v'-t}{\mu} = \frac{\mu v}{v}, \text{ when } t \text{ is negligible.}$$

Hence the resultant magnification

$$m = m_1 \times m_2 = \frac{v'}{\mu u} \times \frac{\mu v}{v'} = \frac{v}{u}.$$

(b) *Longitudinal or Axial Magnification* : It is defined as the ratio of the depth of the image to the corresponding depth of the object. Since all objects are essentially three dimensional, it is desirable to study the effect of longitudinal magnification. Denoting it so that differentiating this relation, we have

$$m' = \frac{\delta v}{\delta u},$$

where $$\frac{1}{v} - \frac{1}{u} = \frac{1}{f},$$

so that differentiating this relation, we have

$$-\frac{\delta v}{v^2} - \left(-\frac{\delta u}{u^2}\right) = 0,$$

or $$\frac{\delta v}{v^2} - \frac{\delta u}{u^2} = 0,$$

and therefore $$m' = \frac{\delta v}{\delta u} = \frac{v^2}{u^2} = \left(\frac{v}{u}\right)^2 = m^2 \qquad ...(11)$$

(c) *Angular Magnification* : It is defined as the ratio $\dfrac{\tan \alpha_2}{\tan \alpha_1}$ where a_1 and a_2 are the slope angles of the object ray and the image ray respectively. It is denoted by g.

$$\therefore \qquad \gamma = \frac{\tan \alpha_2}{\tan \alpha_1}$$

By Lagrange-Helmholtz equation,

$$\mu_1 h_1 \sin \alpha_1 = \mu_2 h_2 \sin \alpha_2$$

or $$\mu_1 h_1 \tan \alpha_1 = \mu_2 h_2 \tan \alpha_2,$$

[∵ α_1 and α_2 are small]

Since $$\mu_1 = \mu_2$$

$$h_1 \tan \alpha_1 = h_2 \tan \alpha_2.$$

∴ angular magnification

$$\gamma = \frac{\tan \alpha_2}{\tan \alpha_1} = \frac{h_1}{h_2} = \frac{1}{m}, \qquad \text{... (12)}$$

that is, *the angular magnification is reciprocal of the lateral magnification produced by a lens.*

From relations (11) and (12), we have

$$m' \times \gamma = m^2 \times \frac{1}{m} = m,$$

that is, longitudinal magnification × angular magnification = transverse magnification.

Principal Focal Planes

Planes at right angles to the axis through F_1 and F_2 are called principal focal planes. The property of the first principal focal plane is that if rays at incidence on a convergent lens, are diverging from a point P in the first focal plane, they emerge as a parallel beam, and if rays at incidence on a divergent lens, are converging to a point P in the first focal plane, they too, emerge as a parallel beam, the direction of the emergent pencil being given in both cases by the ray PO or OP which suffers no deviation.

Similarly, for the second principal focal plane if a parallel beam of light is incident on a convex lens, on emergence it will meet at a point in the second principal focal plane and in the case of a concave lens a parallel incident beam on emergence will appear to proceed from a point in the second principal focal plane.

Focal Length of a Thin Lens by Format's Principle : Let P be a point object on the principal axis of a thin convex lens and Q its point image. Let us consider the two rays PAQ and $PO_1 O_2Q$ going from P to Q.

By Format's principle ail the rays take a stationary or a maximum or a minimum time. If v_1 and v_2 be the velocities of light in the media of refractive indices μ_1 and μ_2 respectively, we have

$$\frac{PA+AQ}{v_1} = \frac{PO_1+O_2Q}{v_1}+\frac{O_1O_2}{v_2}$$

or $$PA+AQ = PO_1+O_2Q+\frac{v_1}{v_2}O_1O_2$$

But $$\frac{v_1}{v_2} = \frac{\mu_2}{\mu_1}$$

$$\therefore\ \mu_1\left(PA+AQ\right) = \mu_1\left(PO_1+O_2Q\right)+\mu_2O_1O_2$$

or $$\mu_1\left[\left(PO^2+AO^2\right)^{\frac{1}{2}}+\left(OQ^2+AO^2\right)^{\frac{1}{2}}\right]$$

$$= \mu_1\ (PO + OQ) + (\mu_2 - \mu_1)\ (t_1 + t_2).$$

$$\text{or}\ \mu_1\left[PO\left(1+\frac{AO^2}{2PO^2}+....\right)+OQ\left(1+\frac{AQ^2}{2OQ^2}+....\right)\right]$$

$$= \mu_1\ (PO + OQ) + (\mu_2 - \mu_1)\ (t_1 + t_2).$$

$$\text{or}\ \mu_1PO+\mu_1\frac{AO^2}{2PO}+\mu_1OQ+\mu_1\frac{AO^2}{2OQ}$$

$$= \mu_1\mathrm{PO}+\mu_1\mathrm{OQ}+\left(\mu_2-\mu_1\right)\left(\mathrm{t}_1+\mathrm{t}_2\right)$$

Putting $AO = y$, $PO = u$, $OQ = v$, r_1 = radius of curvature of first surface and r_2 = that of second surface, we have

$$\frac{\mu_1 y^2}{2u} + \frac{\mu_1 y^2}{2v} = (\mu_2 - \mu_1)\left(\frac{y^2}{2r_1} + \frac{y^2}{2r_2}\right),$$

Since $t_1 = \frac{y^2}{2r_1}$ and $t_2 = \frac{y^2}{2r_2}$, (*cf*, Sagitta theorem)

$$\therefore \quad \frac{\mu_1}{v} - \frac{\mu_1}{u} = (\mu_2 - \mu_1)\left(\frac{1}{r_1} + \frac{1}{r_2}\right)$$

Putting signs according to our conventions , we have

$$\frac{\mu_1}{v} - \frac{\mu_1}{u} = (\mu_2 - \mu_1)\left(\frac{1}{r_1} - \frac{1}{r_2}\right),$$

or $$\frac{1}{v} - \frac{1}{u} = \left(\frac{\mu_2 - \mu_1}{\mu_1}\right)\left(\frac{1}{r_1} - \frac{1}{r_2}\right),$$

from which the equations

$$\frac{1}{f} = \left(\frac{\mu_2 - \mu_1}{\mu_1}\right)\left(\frac{1}{r_1} - \frac{1}{r_2}\right)$$

and $$\frac{1}{v} - \frac{1}{u} = \frac{1}{f}$$ follow at once.

Example : *Convex lens of crown glass, refractive index 1.50 is placed in (a) water, $\mu = 1.33$ and (b) carbon bisulphide, $\mu = 1.65$. Determine in each case whether the lens behaves as a converging lens or diverging lens and find its focal length. Assume the focal length of the lens in air is 15 cm.*

Let f, f_w, f_c be the focal lengths of the lens in air, in water and in carbon bisulphide respectively, then

$$\frac{1}{f} = (\mu - 1)\left(\frac{1}{r_1} - \frac{1}{r_2}\right),$$

or $$\frac{1}{15} = (1.50-1)\left(\frac{1}{r_1}-\frac{1}{r_2}\right) = 0.5\left(\frac{1}{r_1}-\frac{1}{r_2}\right) \quad ..(i)$$

Also $$\frac{1}{f_w} = \left(\frac{\mu_2-\mu_1}{\mu_1}\right)\left(\frac{1}{r_1}-\frac{1}{r_2}\right),$$

[where $\mu_2 = 1.50$ and $\mu_1 = 1.33$]

whence, $$\frac{1}{f_w} = \left(\frac{1.50-1.33}{1.33}\right)\left(\frac{1}{r_1}-\frac{1}{r_2}\right)$$

$$= \frac{0.17}{1.33}\left(\frac{1}{r_1}-\frac{1}{r_2}\right) \quad ...(ii)$$

and $$\frac{1}{f_c} = \left(\frac{\mu_2-\mu_3}{\mu_3}\right)\left(\frac{1}{r_1}-\frac{1}{r_2}\right)$$

[where $\mu_2 = 1.50$ and $\mu_3 = 1.65$]

$\therefore$ $$\frac{1}{f_c} = \left(\frac{1.50-1.65}{1.65}\right)\left(\frac{1}{r_1}-\frac{1}{r_2}\right)$$

$$= \frac{0.15}{1.65}\left(\frac{1}{r_1}-\frac{1}{r_2}\right) \quad ...(iii)$$

Dividing (*i*) by (*ii*), we get $\frac{f_w}{15} = +0.5 \times \frac{1.33}{0.17}$

or $$f_w = +\frac{0.5 \times 1.33 \times 15}{0.17} = +\ 58.7 \text{ cm.}$$

The negative sign indicates that the lens behaves as a diverging lens.

Dividing (*i*) by (*iii*)

$$\frac{f_e}{15} = -\ \frac{0.5 \times 1.65}{0.15},$$

$$\therefore \qquad f_e = -\frac{0.5 \times 1.65 \times 15}{0.15} = -\ 82.5 \text{ cm.}$$

The negative sign indicates that the lens behaves as a diverging lens.

Example : *A bi-convex lens of radii of curvature 30 0 cm each is placed on the surface of water so that the lower surface is only immersed. An illuminated object 40 cm deep is observed vertically through the lens and water. Find the position of the image. Refractive index of air glass = 3/2 and for air-water = 4/3.*

When there are media of refractive indices μ_1 and μ_2 on either side of the lens of refractive index μ_2 then tor refraction at the first surface

$$\frac{\mu_2}{v'} - \frac{\mu_1}{u} = \frac{\mu_2 - \mu_1}{r_2} \qquad \dots (a)$$

and for refraction at the second surface

$$\frac{\mu_2}{v} - \frac{\mu_2}{v'} = \frac{\mu_3 - \mu_2}{r_2} \qquad \dots (b)$$

neglecting the thickness of the lens.

Adding (*a*) and (*b*)

$$\frac{\mu_3}{v} - \frac{\mu_2}{u} = \frac{\mu_3 - \mu_1}{r_1} + \frac{\mu_3 - \mu_2}{r_2} \qquad \dots (c)$$

Here, $\qquad \mu_1 = \frac{4}{3}, \ \mu_2 = \frac{3}{2}, \ \mu_3 = 1.0$

Conventionally $u = -40$ cm, $r_1 = 30$ cm and $r_2 = -30$ cm.

Substituting the values in equation (*c*), we have

$$\frac{1.0}{v} - \frac{\frac{4}{3}}{-40} = \frac{\frac{3}{2} - \frac{4}{3}}{30} + \frac{1.0 - \frac{3}{2}}{-30}$$

or $$\frac{1}{v}+\frac{1}{30} = \frac{1}{180}+\frac{1}{60},$$

or $$\frac{1}{v} = \frac{1}{180}+\frac{1}{60}-\frac{1}{30}$$

$$= \frac{1+3-6}{180} = -\frac{2}{180},$$

whence $v = -\ 90$ cm

That is, the image will be on the same side as the object at a distance of 90 cm from the lens and is virtual.

Example : *A convex lens of focal length 20 cm and made of glass (μ = 1.50) is immersed in water (μ = 1.33). Find the change in the focal length of the lens.*

For lens in air,

$$\frac{1}{f} = (\mu-1)\left(\frac{1}{r_1}-\frac{1}{r_2}\right),$$

where $\mu = 1.50$ and $f = 20$ cm.

$$\therefore \quad \frac{1}{20} = (1.50-1)\left(\frac{1}{r_1}-\frac{1}{r_2}\right) = 0.5\left(\frac{1}{r_1}-\frac{1}{r_2}\right), \quad \ldots (i)$$

For lens in water,

$$\frac{1}{f_1} = \left(\frac{1.50-1.33}{1.33}\right)\left(\frac{1}{r_1}-\frac{1}{r_2}\right)$$

$$= \frac{0.17}{1.33}\left(\frac{1}{r_1}-\frac{1}{r_2}\right) \quad \ldots (iii)$$

Dividing (*i*) by (*ii*)

$$\frac{f_1}{20} = \frac{0.5\times 1.33}{0.17},$$

or $$f_1 = \frac{0.5\times 1.33\times 20}{0.17} = +\ 78.2 \text{ cm}.$$

The positive sign indicates that the lens behaves as a

converging lens of focal length 78.2 cm so that the increase in focal length is equal to 78.2 – 20 = 58.2 cm.

Example : *A plano-convex lens of focal length 15 cm is fixed at the centre of one side of a box with convex surface outside. An object 5 cm high is placed at 25 cm from the lens. Find the position and size of the image, when the box is (a) empty, (b) filled with water.*

$${}^a\mu_g = \frac{3}{2};\ {}^a\mu_w = \frac{4}{3}.$$

(*a*) When the box is empty,

$$\frac{1}{v} - \frac{1}{u} = \frac{1}{f},$$

or $$\frac{1}{v} - \frac{1}{-25} = \frac{1}{15}, \qquad \text{whence } v = 37.5 \text{ cm.}$$

Now, $$\frac{\text{size of image}}{\text{size of object}} = \frac{v}{u} = \frac{37.5}{25}$$

whence, size of image $= \dfrac{37.5 \times 5}{25} = 7.5$ cm.

Since the lens is plano-convex with convex surface outside

$$\therefore \qquad \frac{1}{f} = (\mu - 1)\left(\frac{1}{r_1} - \frac{1}{\infty}\right)$$

or $$\frac{1}{15} = \frac{0.5}{r_1}, \text{ whence } r_1 = \frac{15}{2} \text{ cm:}$$

(*b*) When there is water on the other side, with usual notation, we have

$$\frac{\mu_3}{v} - \frac{\mu_1}{u} = \frac{\mu_2 - \mu_1}{r_1} - \frac{\mu_3 - \mu_2}{r_2}$$

Substituting the values in this equation, we get

$$\frac{\frac{4}{3}}{v} - \frac{1}{-25} = \frac{\frac{3}{2} - 1}{\frac{15}{2}} - \frac{\frac{4}{3} - \frac{3}{2}}{\infty} = \frac{1}{15},$$

$$\text{or} \qquad \frac{4}{3v} = \frac{1}{15} - \frac{1}{25} = \frac{2}{75},$$

$$\text{or} \qquad v = 50 \text{ cm}$$

$$\text{Size of image} = \frac{50}{25} \times 5 = 10 \text{ cm.}$$

Deviation Method

Derivation of Lens Formula by Deviation Method : This method is based on the prismatic properties of a lens, which may be regarded as a series of prisms of small angles which increase in size of the refracting angles from the centre of the lens outwards. In the case of a convex lens these refracting angles are pointing away from the axis and in the case of a concave lens, they are pointing inwards. We know that for a prism of small angle a, the deviation *S*, is give by

$$\delta = (\mu - 1)\ \alpha,$$

where μ is the refractive index of the material of the prism.

In deriving the lens formula we shall make use of the paraxial rays and assume that the medium on both sides of the lens is the same unless otherwise stated.

If a ray *PA* parallel to the axis is incident at a point *A* distant *h* from the centre of the lens, on emerging it meets the axis in *F* and the deviation produced is given by

$$\delta = \frac{h}{f},$$

where f is the focal length.

Next, let *P* be a point object lying on the principal axis. A ray *PA* incident at the same distance *h* from the centre of the lens suffers the deviation δ_1 and after refraction cuts the axis in *Q*. Since the deviation is independent of the angle of incidence- for paraxial rays, $\delta_1 = \delta$.

From the geometry of,

$$\delta = \frac{h}{f},$$

and $$\delta_1 = \alpha + \beta = \frac{h}{-u} + \frac{h}{v}$$

$\therefore$ $$\frac{h}{f} = \frac{h}{-u} + \frac{h}{v}$$

or $$\frac{1}{f} = \frac{1}{v} - \frac{1}{u}.$$

The fundamental lens formula can be derived as under:

We have seen above that the angle of deviation is given by

$$\delta = \frac{h}{f},$$

where h is the height of the point of incidence above the axis.

Now, $$\delta = (\mu - 1)\ \alpha,$$

where α is the refracting angle of the prism forming the corresponding portion of the lens α, therefore, is the angle between the tangent planes to the lens faces at the distance h from the axis.

We find that a is also the angle between the radial drawn to the lens faces at the same distance.

so that $$\alpha = \theta + \phi,$$

whence $$\delta = (\mu - 1)\ (\theta + \phi).$$

But $$\delta = \frac{h}{f}, \text{ therefore } \frac{h}{f} = (\mu - 1)\ (\theta + \phi)$$

Now, conventionally $\theta = \dfrac{h}{r_1}$ and $\phi = -\dfrac{h}{r_2}$.

$$\therefore \qquad \frac{h}{f} = (\mu - 1)\left(\frac{h}{r_1} - \frac{h}{r_2}\right),$$

or
$$\frac{1}{f} = (\mu - 1)\left(\frac{1}{r_1} - \frac{1}{r_2}\right).$$

Example : *A thin air double concave lens bounded by very thin glass walls with radii of curvature 24 cm for each surface separates a long tank into two compartments one of which is filled with water of refractive index 1.33 and the other with a liquid of refractive index 1.67. If a parallel beam of light traversing the water strikes the lens, find where it is focussed ?*

For refraction at the first surface,

$$\frac{\mu_2}{v'} - \frac{\mu_1}{u} = \frac{\mu_2 - \mu_1}{r_1}.$$

For refraction at second surface,

$$\frac{\mu_2}{v} - \frac{\mu_2}{v'} = \frac{\mu_3 - \mu_2}{r_2}.$$

Adding
$$\frac{\mu_3}{v} - \frac{\mu_1}{u} = \frac{\mu_2 - \mu_1}{r_1} + \frac{\mu_3 - \mu_2}{r_2}.$$

Since incident light is a parallel beam, $u = -\infty$, $v = f$,

$$\therefore \qquad \frac{\mu_3}{f} = \frac{\mu_2 - \mu_1}{r_1} + \frac{\mu_3 - \mu_2}{r_2}.$$

Here, $\mu_3 = 1.67$; $\mu_2 = 1.00$; $\mu_1 = 1.33$; $r_1 = -24$; $r_2 = +24$.

$$\therefore \qquad \frac{1.67}{f} = \frac{1.00 - 1.33}{-24} + \frac{1.67 - 1.00}{24}$$

$$= \frac{0.33}{24} + \frac{0.67}{24} = \frac{1}{24}$$

whence
$$f = 1.67 \times 24 = 40.1 \text{ cm}.$$

Hence the image is 40.1 cm from the lens in the second liquid.

Minimum Distance between the Object and its Real Image in a Convex Lens : In order to obtain the real image of an object on the screen by means of a convex lens the object and the screen must be placed at a distance apart which is greater than a certain minimum value. Let x be this distance. Then,

$$x = OP + OQ = -u + v \qquad ...(i)$$

and we have to determine the minimum value of x subject to the condition that

$$\frac{1}{v} - \frac{1}{u} = \frac{1}{f} \qquad ... (ii)$$

Eliminating u from these equations, we have

$$\frac{1}{v} - \frac{1}{v-x} = \frac{1}{f}$$

or $$\frac{v-x-v}{v(v-x)} = \frac{1}{f}$$

or $$v(v-x) = -xf$$

or $$v^2 - vx + xf = 0$$

Solving the quadratic for v, we get,

$$v = \frac{x \pm \sqrt{x^2 - 4xf}}{2} \qquad ... (iii)$$

For v to be real

$$x^2 - 4xf = + \text{ve or zero.}$$

$\therefore$ $$x^2 \geqq 4\,xf$$

or $$x \geqq 4f.$$

That is, the *minimum* value of x is $4f$.

or *alternatively*, we may use the method of differential calculous.

Again $$x = v - u$$

and $$\frac{1}{v} - \frac{1}{u} = \frac{1}{f}.$$

Eliminating v from these equations, we have

$$x = \frac{uf}{u+f} - u.$$

Differentiating, $$\frac{dx}{du} = \frac{(u+f)f - uf}{(u+f)^2} - 1 = \frac{f^2 - (u+f)^2}{(u+f)^2}$$

$$= \frac{-u(u+2f)}{(u+f)^2}.$$

Now for x to be a minimum or a maximum

$$\frac{dx}{du} = 0,$$

so that, $u = 0,$

or $u + 2f = 0.$

The value $u = 0$ is of no physical consequence, as it means that the object is in contact with the lens. The second value $u = -2f$, when substituted in

$$\frac{1}{v} - \frac{1}{u} = \frac{1}{f}, \text{ gives } v = 2f.$$

Then, $x = 2f - (-2f) = 4f.$

Differentiating again and simplifying, we get

$$\frac{d^2x}{du^2} = \frac{-2f^2}{(u+f)^2}$$

Substituting $u = 2f$, gives

$$\frac{d^2x}{du^2} = \frac{2}{f}, \text{ a positive value, since } f \text{ is + ve.}$$

Therefore, the value of $x = 4f$ is *minimum* under these conditions.

Note : From the relation (*iii*) above, it is obvious that there are two values of v given by

$$v_1 = \frac{x + \sqrt{x^2 - 4xf}}{2}$$

and
$$v_2 = \frac{x - \sqrt{x^2 - 4xf}}{2}$$

Since, $u = v - x$, we have

$$u_1 = v_1 - x = \frac{-x + \sqrt{x^2 - 4xf}}{2}$$

and
$$u_2 = v_2 - x = \frac{-x - \sqrt{x^2 - 4xf}}{2}$$

Thus, we have

$$u_1 = -v_2 \text{ and } u_2 = -v_1.$$

Hence, in general there are two coaxial positions of a convergent lens which will produce on a fixed screen, a sharp image of a fixed object.

Two Thin Lenses in Contact : Sometimes for removing certain aberrations we have to make use of two lenses placed in contact. We shall therefore, find the combined focal length of such a lens system. Let two thin lenses of focal lengths f_1 and f_2, be placed in contact so that both have a common axis. An object at a distance u from the first lens of focal length f_1, gives an image at a distance v' from it, so that

$$\frac{1}{v'} - \frac{1}{u} = \frac{1}{f_1} \qquad \text{... } (a)$$

This image serves as a virtual object for the second lens and since the lenses are thin and in contact the final image is formed at a distance v from the lens system given by

$$\frac{1}{v} - \frac{1}{v'} = \frac{1}{f_2} \qquad \text{... } (b)$$

Adding (a) and (*b*), v' is eliminated and we get

$$\frac{1}{v} - \frac{1}{u} = \frac{1}{f_1} + \frac{1}{f_2}.$$

But $$\frac{1}{v} - \frac{1}{u} = \frac{1}{F}$$

where F is the focal length of the combination.

$$\frac{1}{F} = \frac{1}{f_1} + \frac{1}{f_2}. \quad \text{... (14)}$$

In other words, the reciprocal of the focal length of a thin lens combination is equal to the sum of the reciprocals of the focal lengths of the component lenses.

The single lens of focal length F, when placed in the position occupied by the lens combination produces an image of a given object in the same position and of the same size as that produced by the combination.

Since power of a lens, $P = 1/f$, we obtain for the power of the combination

$$P = P_1 + P_2. \quad \text{... (15)}$$

So that when two thin lenses are placed in contact, *the power of the combination is equal to the sum of the powers of the component lenses.*

In general, if there are n thin lenses, we have

$$\frac{1}{F} = \frac{1}{f_1} + \frac{1}{f_2} + \frac{1}{f_3} + \dots\dots\dots + \frac{1}{f_n} \quad \text{... (16)}$$

and $$P = P_1 + P_2 + \dots\dots + P_n. \quad \text{... (17)}$$

The relations (14), (15), (16) and (17) hold good whether, the lenses are all convex or all concave or some convex and some concave.

7

Definite Laws

When light travelling in a homogeneous medium is incident on a second medium, some of it is sent back into the first medium according to definite laws and is said to be reflected.

The angles which the incident ray and the reflected ray make with the normal to the surface are termed the angles of incidence and of reflection respectively. The laws of reflection are as follows:

Law 1. The incident ray, the reflected ray and the normal to the surface, all lie in the same plane.

Law 2. The angle of incidence i is equal to the angle of reflection r.

The deviation suffered by the reflected ray is the angle by which its path has been deflected. It is equal to $180 - (i + r)$. Since i and r are both equal, the deviation is given by $D = 180 - 2i$.

The student is supposed to be familiar with some of the consequences of reflection at plane surfaces, so we shall consider in the following sections the formation of images by reflection at curved faces only.

Spherical Mirrors

Let us first of all assume that the surfaces are portions of a sphere. If the inside surface reflects, the mirror is concave and if it is the outside surface which reflects, it is convex.

The aperture of the mirror is the circular face presented to the incident tight and the width of the aperture is called its diameter. The centre of the aperture is called the pole of the mirror and the centre of the sphere of which the mirror forms a part is termed the centre of curvature. The line joining, the centre of curvature and the pole of the mirror is the principal axis. The laws of reflection stated above also hold in the case of spherical mirrors. However, before we proceed to determine the algebraic relations between the radius of curvature of a mirror, the distances of the object and the image from its pole we shall mention the convention of signs for distances and angles adopted in this book.

Convention of Signs. In deducing algebraic relations and expressions connecting distances of objects and images in all cases of reflection and refraction at spherical surfaces and in cases of lenses too, both converging and diverging, it is desirable to have a single formula applicable to both kinds of mirrors or surfaces or lenses, so it is necessary to adopt some sign conventions for *distances* and *angles* measured. Different writers have used different sign conventions, for each serves the purpose as well as any other. But to avoid confusion arising thereof the sign conventions which are now being largely used by most of the writers are those recommended by the International Commission on Optics and the same will be adopted in the following chapters on Geometrical Optics and the student is advised to keep them firmly in mind.

1. Distances measured from the centre of mirror surface or lens in the direction in which light is travelling are counted as 'positive' while those measured in the opposite direction are taken as 'negative'.

2. Transverse distances are counted as 'positive' when measured upwards from principal axis and 'negative' when measured below it.

3. The angles of incidence and reflection or refraction are taken as 'positive' when the radius of the surface must be rotated counter-clockwise through an angle less than $\pi/2$ to coincide with the direction of the ray.

4. The slope angles, *viz.*, the angles which the incident and reflected or refracted rays make with the axis are 'positive' when the axis of the surface must be rotated counter- clockwise through an angle less than $\pi/2$ to coincide with the direction of the ray.

It may be noted that these conventions agree with the usual Coordinate Geometry conventions for coordinates and positions so that if the centre of the mirror or lens or the curved surface is placed at the imaginary origin of the coordinates and the axis of the optical system coincides with the X-axis, the distances on the right of the origin are positive and on the left negative. The student is already familiar with the coordinate geometry conventions and therefore he shall have no difficulty in applying them readily to mirrors and lenses.

Mirror Formula

Let *AMB* be the principal section of a spherical mirror with a small aperture and centre of curvature at

C. Let a ray *PA*, proceeding from a point source *P* lying off the axis, be reflected at *A* so that it makes equal angles with the normal *CA*, *i.e.*, $\angle i = \angle r$. The ray *PM* incident normally on the mirror reflects back on itself. These two rays intersect on the axis to form image Q. The mirror is concave and convex to incident light. Let the angles which PA, QA and CA make with the principal axis be denoted by α, β and γ respectively. Then for the concave mirror

$$i = \gamma - \alpha$$

and $$r = \beta - \gamma$$

since $$i = r$$

$$\therefore \quad \beta - \gamma = \gamma - \alpha,$$

or $$\beta + \alpha = 2\gamma. \qquad \text{... } (a)$$

Now $$\alpha = \frac{AM}{MP} = \frac{AM}{MQ}$$

and $$\gamma = \frac{AM}{MC}$$

If we denote AM, MP, MQ and MC by h, $-u$, $-v$ and $-r$ respectively, due regard being paid to the convention of signs,

then $$\alpha = \frac{h}{-u}, \ \beta = \frac{h}{-v}$$

and $$\gamma = \frac{h}{-r}$$

and equation (a) may be written as

$$\frac{h}{-v} + \frac{h}{-u} = \frac{2h}{-r}$$

or $$\frac{1}{v} + \frac{1}{u} = \frac{2}{r}.$$

For the convex mirror

$$i = \alpha + \gamma$$

and $\qquad r = \beta - \gamma.$

Since $\qquad i = r, \quad \therefore \alpha + \gamma = \beta - \gamma,$

or $\qquad \beta - \alpha = 2\gamma. \qquad \ldots (b)$

Now $$\alpha = \frac{AM}{MP} = \frac{h}{-u};$$

$$\beta = \frac{AM}{MQ} = \frac{h}{v}$$

and $$\gamma = \frac{AM}{MC} = \frac{h}{r}.$$

Putting these values in equation (*b*) above, we have

$$\frac{h}{v} - \frac{h}{-u} = \frac{2h}{r},$$

or $$\frac{1}{v} + \frac{1}{u} = \frac{2}{r}. \qquad \ldots (1)$$

This is the same result as obtained above for a concave mirror.

It may be noted that this equation is independent of the angles of incidence, provided the aperture is *small,* so that the position of *Q* is the same for the given position of *P* for all values of the angle of incidence.

If the object is at infinite distance from the mirror, the particular point at which image is formed is called the focus of the mirror and the distance of the point from *M* is called the focal length of the mirror, *i.e.,*

$$f = \underset{u \to \infty}{\text{Lt.}} [v]$$

Substituting in (1), we get

$$\frac{1}{f} + \frac{1}{\infty} = \frac{2}{r} \text{ or } f = \frac{r}{2}.,$$

that is, focal length is equal to half the radius of curvature, the formula (1) thus becomes

$$\frac{1}{v}+\frac{1}{u} = \frac{1}{f} \qquad \ldots (1a)$$

The points P and Q connected by this relation are known as conjugate points and transverse planes through these points are called conjugate planes.

Important note. *In using the relation*

$$\frac{1}{v}+\frac{1}{u} = \frac{1}{f}$$

in the solution of numerical problems, the student must never forget to substitute in it the conventional signs for the quantities whose values are given. No sign is to be attached to the quantity to be determined.

Focal plane. A plane perpendicular to the axis and drawn through the focus is called focal plane. Incident parallel rays making some angle with the principal axis are brought to focus at some point in the focal plane.

The image Q of a distant point object, sending an oblique pencil of parallel rays is formed at the intersection with focal plane of that ray which goes through the centre of curvature.

Spherical Aberration

In deducing the general formulae for mirrors, we assumed that the angles between the incident rays or the reflected rays and the principal axis are small and a bundle of parallel rays close to the axis are brought to a sharp focus. This assumption is not justified when the aperture of the mirror is large. Rays that are reflected from a limited region of the mirror in the neighbourhood of the vertex M, are brought to focus at a point farther from the pole than marginal rays which cross the axis at points nearer the mirror.

This failure of rays to converge to or appear to diverge from a single point after reflection from a spherical mirror is called 'spherical aberration'. The effect of this is to destroy the sharpness of the image that would otherwise be formed by the mirror. Each ray after reflection is seen to converge and cut its neighbouring reflected ray at a point which lies on a curve called the caustic curve, the focus of the paraxial rays forming the cusp of the caustic.

We often see such a curve on the surface of tea in a cup, the light being reflected from the inside of the cup. It is the curve of greatest luminosity with its apex at F, the principal focus. The distance between the foci for the paraxial and marginal rays is taken as a measure of the axial spherical aberration. The rays reflected from the mirror are only drawn in one plane, *i.e.*, the plane of the paper.

Since the mirror is a part of a sphere, the caustic curve is an intersection by the plane of the paper of a surface which the reflected rays from all parts of the mirror touch. This surface is known as caustic surface. It is the surface traced out by the caustic curve; about the principal axis.

If a screen is placed at the paraxial focus F and then shifted towards the mirror, a point is reached where the size of the circular image is the *least*. This circular spot is called the circle of least confusion and may be taken as the nearest approach to the image of the distant object.

Aberration by reflection at a concave surface. Let P be a point object lying on the axis MC of a concave mirror, with its centre of curvature at C and Q' be its image formed by reflection of a ray PA incident on the mirror at a point for removed from the axis. Let $\angle ACM = \gamma$, $PM = u$, $Q'M = v'$ $CM = r$ and $AN = h$ and it is such that h^3 and higher powers may be neglected in comparison with corresponding powers of u etc.

Since AC bisects the angle PAQ'.

$$\frac{Q'C}{CP} = \frac{AQ'}{AP}$$

or $\quad Q'C.AP = CP \,.\, AQ' \quad \ldots (1)$

Now $\quad AP^2 = AC^2 + CP^2 - 2AC \,.\, CP \,.\, \cos(180 - \gamma)$

[cosine formula]

$$= AC^2 + CP^2 + 2AC\,.\,CP\cos\gamma$$

$$= (-r)^2 + (-u+r)^2 + 2\times(-r)(-u+r)\cos\gamma\,.$$

$$\because \quad \cos\gamma = \sqrt{1-\sin^2\gamma}$$

$$= \left(1-\frac{h^2}{r^2}\right)^{\frac{1}{2}} = 1-\frac{h^2}{2r^2} \text{ approx.,}$$

$$\therefore \quad AP^2 = r^2 + (u-r)^2 + 2r(u-r)\left(1-\frac{h^2}{2r^2}\right)$$

$$= u^2 - \frac{u-r}{r}h^2 \qquad \text{on simplification}$$

$$= u^2\left[1-\frac{u-r}{ur}\cdot\frac{h^2}{u}\right]$$

$$= u^2\left[1-\left(\frac{1}{r}-\frac{1}{u}\right)\cdot\frac{h^2}{u}\right]$$

$$\therefore \quad AP = u\left[1-\left(\frac{1}{r}-\frac{1}{u}\right)\cdot\frac{h^2}{u}\right]^{\frac{1}{2}}$$

$$= u\left[1-\left(\frac{1}{r}-\frac{1}{u}\right)\cdot\frac{h^2}{2u}\right]$$

by binomial expansion.

Similarly, $AQ' = v'\left[1-\left(\frac{1}{r}-\frac{1}{v'}\right)\cdot\frac{h^2}{2v'}\right]$

Also $\quad Q'C = (r - v')$

and $\quad CP = (u - r)$.

Substituting the values in equation (*i*), we get

$$(r-v')u\left[1-\left(\frac{1}{r}-\frac{1}{u}\right)\cdot\frac{h^2}{2u}\right] = (u-r)v'\left[1-\left(\frac{1}{r}-\frac{1}{v'}\right)\cdot\frac{h^2}{2v'}\right]$$

Dividing throughout by $uv'r$, we have

$$\left(\frac{1}{v'}-\frac{1}{r}\right)\left[1-\left(\frac{1}{r}-\frac{1}{u}\right)\cdot\frac{h^2}{2u}\right] = \left(\frac{1}{r}-\frac{1}{u}\right)\left[1-\left(\frac{1}{r}-\frac{1}{v'}\right)\cdot\frac{h^2}{2v'}\right]$$

Removing the brackets and simplifying, we get

$$\frac{1}{v'}+\frac{1}{u} = \frac{2}{r}+\left(\frac{1}{r}-\frac{1}{u}\right)\left(\frac{1}{v'}+\frac{1}{r}\right)\left(\frac{1}{v'}+\frac{1}{u}\right)\frac{h^2}{2} \quad ..\ (ii)$$

Since h^2 is small we can substitute v for v' in the terms forming the coefficient of h^2, where v is the image distance for paraxial rays.

$$\therefore \quad \frac{1}{v'}+\frac{1}{u} = \frac{1}{v}+\frac{1}{u} = \frac{2}{r}$$

and the relation

$$\frac{1}{v'}+\frac{1}{u} = \frac{2}{r} \text{ may be written as}$$

$$\frac{1}{v'}-\frac{1}{r} = \frac{1}{r}-\frac{1}{u}$$

Therefore in the expression on the right-hand side of equation (*ii*) we can substitute.

$\left(\frac{1}{r}-\frac{1}{u}\right)^2$ for $\left(\frac{1}{r}-\frac{1}{u}\right)\left(\frac{1}{v'}-\frac{1}{r}\right)$ and $\frac{2}{r}$ for $\frac{1}{v'}+\frac{1}{u}$.

Hence equation (*ii*) becomes

$$\frac{1}{v'}+\frac{1}{u} = \frac{2}{r}+\left(\frac{1}{r}-\frac{1}{u}\right)^2 \frac{h^2}{r} \quad \text{... (iii)}$$

$$\text{or} \quad \frac{1}{v'}-\left(\frac{2}{r}-\frac{1}{u}\right) = \left(\frac{1}{r}-\frac{1}{u}\right)^2 \frac{h^2}{r},$$

$$\text{or} \quad \frac{1}{v'}-\frac{1}{v} = \left(\frac{1}{r}-\frac{1}{u}\right)^2 \frac{h^2}{r} \quad \left[\because \frac{2}{\mathrm{r}}-\frac{1}{\mathrm{u}}=\frac{1}{\mathrm{v}}.\right]$$

$$\therefore \quad v'-v = \left(\frac{1}{r}-\frac{1}{u}\right)^2 \frac{h^2}{r} v^2 \quad [\because \mathrm{v'} \to \mathrm{v}.]$$

If we put $h = a$, the semi-aperture of the mirror, we get

$$v'-v = \left(\frac{1}{r}-\frac{1}{u}\right)^2 \frac{a^2}{r} v^2$$

$$\text{or} \quad v'-v = \frac{\left(\frac{1}{r}-\frac{1}{u}\right)^2}{\frac{1}{v^2}} \times \frac{a^2}{r}.$$

$$\text{But} \quad \frac{1}{v} = \frac{2}{r}-\frac{1}{u}$$

$$\therefore \quad v'-v = \frac{\left(\frac{1}{r}-\frac{1}{u}\right)^2}{\left(\frac{2}{r}-\frac{1}{u}\right)^2} \times \frac{a^2}{r} \quad \text{... (2)}$$

Thus knowing u, r and a, $(v'-v)$ can be calculated.

The expression giving the value of $(v'-v)$ is a measure of the axial spherical aberration.

Astigmatism

If a narrow cone of light rays is incident *obliquely* on a small portion of a large concave mirror the reflected rays do not form a sharp focus even if the aperture is small. They are concentrated into a pair of broadened and mutually perpendicular lines called focal lines. This effect is known as astigmatism and the reflected pencil of rays is called astigmatic-pencil. Consider two rays *PA* and *PB* in the meridional or vertical plane diverging from a point *P* on the axis and incident on a small portion *AB* of the mirror making a large angle with the principal axis *PM*. Let these rays after reflection cut each other in *T* and meet the principal axis in *Q* and *S* respectively. When *A* and *B* are very close together, *T* assumes a definite limiting position, *viz.*, the so-called *meridional focus.* Next consider the pencil with its principal ray *PA* proceeding along *AQ* after reflection. A neighbouring ray in the sagittal plane of the pencil would meet the mirror in *A′*, a point near *A* and would after reflection pass through *Q*, since *PA′C* is the plane containing the reflected ray. Then *Q* is the *sagittal focus.* The ΔPAB traces out a pencil of rays diverging from *P*. The point *T* traces out a short line *TT′* as shown perpendicular to the paper and the reflected rays pass through this line and also through the line *SQ* along the axis. The line *TT′* is called the first focal line F_1, and *SQ* the second focal line F_2 the light in this focal region forming a sort of *lemniscate.*

At a certain place in between the focal lines the length and the breadth of the patch of light are equal. This place is called the *circle of least confusion,* being the nearest approach to an image formed by the reflected pencil. The distance between the two focal lines is called astigmatic difference.

Expression for astigmatic difference. Where *Q* it the sagittal focus.

Let $$AP = p,$$
$$AQ = q$$

and $$AC = r,$$

$\therefore$ $$\Delta CAQ + \Delta PAC = \Delta\ PAQ$$

we have,

$$rq \sin\theta + rp \sin\theta = pq \sin 2\theta,$$

or $$rq \sin\theta + rp \sin\theta = 2pq \sin\theta \cos\theta.$$

Dividing by $pqr \sin\theta$, we have

$$\frac{1}{p} + \frac{1}{q} = \frac{2\cos\theta}{r} \qquad \text{...(3)}$$

T is the meridional focus, if α, β and γ are the angles as indicated and $AT = q'$ then

$$\theta + \beta = \theta + \delta\theta + \gamma,$$

or $$\delta\theta = \beta - \gamma$$

and $$\theta + \gamma = \theta + \delta\theta + \alpha,$$

or $$\delta\theta = \gamma - \alpha,$$

$\therefore$ $$\beta - \gamma = \gamma - \alpha$$

or $$\alpha + \beta = 2\gamma. \qquad \text{... }(a)$$

Now, $$\alpha = \frac{AB\cos\theta}{p},$$

$$\beta = \frac{AB\cos\theta}{q'},$$

$$\gamma = \frac{AB}{r}.$$

Putting these values in equation (a), we get

$$\frac{AB\cos\theta}{p} + \frac{AB\cos\theta}{q'} = \frac{2AB}{r},$$

whence $$\frac{1}{p} + \frac{1}{q'} = \frac{2}{r\cos\theta}. \qquad \text{... (4)}$$

Subtracting equation (3) from equation (4), we get

$$\frac{1}{q'} - \frac{1}{q} = \frac{2}{r}\left[\frac{1}{\cos\theta} - \cos\theta\right]$$

$$= \frac{2}{r}\frac{1-\cos^2\theta}{\cos\theta} = \frac{2\sin^2\theta}{r\cos\theta}$$

$$= \frac{2}{r}\cos\theta\tan\theta,$$

or, the astigmatic difference

$$q - q' = \frac{2qq'}{r}\sin\theta\tan\theta \qquad ...(5)$$

Obviously, therefore, the astigmatic difference rapidly increases with increasing angle of incidence.

Note. Spherical aberration and Astigmatism can be reduced to a large extent by the use of a stop at the proper place so that the beam is incident at the central part of the mirror only.

The astigmatic effect for a parallel beam of rays originating in a point object at infinity and making a large angle with the principal axis *CM* of a concave mirror having a small aperture *AB*, centre C and polo *M*. Here again as before the rays give rise to two focal lines F_1 and F_2 separated by a finite distance and mutually perpendicular, so that even a mirror of small aperture will not produce a sharply focused image of a point object remote from its axis

From the above discussion we conclude that for oblique incidence or for wide-angled pencils a spherical mirror acts as if focal length were less than for direct pencils, the result being that a large object will give a curved image, which is also distorted, different parts being disproportionately magnified. Moreover, for points near the edge of the object large circles of confusion are produced and these cause considerable blurring of the image.

Ellipsoidal and Paraboloidal Mirrors

To reduce spherical aberration we make use of ellipsoidal and paraboloidal mirrors. *CXD* is a section of the ellipsoidal mirror. The surface of such a mirror is one which is generated by an ellipse revolving about its major axis. The rays proceeding from a point source placed at one of the foci of the ellipse, say *P*, from, after reflection, a point image at the other focus *P′* no matter where they meet the mirror, the total length of the path of each ray between *P* and *P′* being always the same. The incident ray *PA* and the reflected ray *AP′* make equal angles with the normal drawn at *A* and so is the case with any other ray. Such a reflecting surface is called an aplanatic surface and is defined by the equation $PA + AP'$ =constant. The points *P* and *P′* are called aplanatic foci.

When one of the foci, either *P* or *P′* is at infinite distance as the aplanatic surface for reflection becomes a paraboloid of revolution. It is a more commonly used form of the mirror in which incident rays, parallel to the axis after reflection, are all concentrated at a single point as a consequence of the geometrical property of the parabola, *viz.*, the incident ray *AB*, parallel to the axis and the reflected ray *BS* are equally inclined to the normal at *B*, wherever the point *B* may be. There is, thus, a complete absence of spherical aberration, no matter how large the aperture of the mirror. This property has commended its use in the construction of high resolving power concave reflectors of the modern telescopes. The Mount Palomar telescope has a concave parapodial reflector with 5.08 m (200 inch) aperture.

Conversely, it may also be noted that a point source of light at the focus of a paraboloidal mirror will send out parallel rays after reflection from the surface of the mirror. It is for this reason that such mirrors are used in automobile headlights and searchlights.

Bibliography

Addicott, F.T. : *Optics,* University of California Press, London, 1982.

Atherton, J.C. : *Manipulation of Flowering,* Butterworths, London, 1987.

Ball. N.G., M.B. Wilkins : *The Physiology of Plant Growth and Development,* McGraw Hill book, London, 1969.

Bemier, G., J.M. Kinet : *Physical Optics,* C.R.C. Press, Boca Raton, 1981.

Benjamin, W.A. : California. Boyer, P.D. : *The Enzymes,* Academic Press, New York, 1970.

Bull, H.B. : *An Introduction to Physical Optics,* F.A. Davis Co. Philadelphia, Eisenburg, 1964.

Canny, M.J. : *Phloem Translocation,* Cambridge University Press, New York, 1973.

Cohen, P. : *Control of Enzyme Activity,* Chapman and Hall, London, 1976.

Crafts, A.S. and C.E. Crisp, : *Phloem Transport in Plants,* W.H. Freeman, San Francisco, 1973.

Douglas, J.S. : *Advanced Guide to Optics,* Pelham Books, London, 1976.

Epstein, E. : *Geometrical Optics,* John Wiley, New York, 1972.

Epstein, E. : *Mineral Nutrition of Plants : Principles and Perspectives,* John Wiley and Sons, New York, 1972.

Esau, K. : *Plant Anatomy,* John Wiley and Sons, New York, 1965.

Fershl, A. : *Enzyme Structure and Mechanism,* W.H. Freeman, San Francisco, 1977.

Forward, D.F. : *Energy and Carbon Metabolism,* Academic Press, New York. 1983.

Ftzler M.E. : *Plant Lectins : Molecular and Biological Aspects,* Academic Press, New York, 1985.

Gauch, H.G. : *Inorganic Plant Nutritionm,* Dowden, Hutchinson and Ross, Pennsylvania, 1972.

Govindjee, : *Bioenergetics of Photosynthesis,* Academic Press, New York, 1975.

Green, J.R. : *An Introduction to Vegetable Physiology,* London, 1911.

Halevy, A.H. : *Handbook of Optics,* C.R.C. Press, Boca Raton, 1985.

Loewy, A.G. and Siekevitz, P. : *Cell Structure and Function,* Rinehart and Winston, New York, 1969.

Mazliak, P. : *Light and Optics,* Chapmann and Hall, London, 1973.

McNeil, M. Darvill : *Structure and Function of the Primary Cell Walls of Plants,* McGraw, London, 1984.

Novikoff, A.B. : *The Endoplasmic Reticulum : a Cytochemist's View,* New York, 1976.

Ogawa, Y. and R.W. King : *Geometrical Optics,* Springer-Verlag, Berlin, 1990.

Opik, H. : *The Respiration of Higher Plants,* Edward Arnold Publishers, London, 1980.

Rabinowitch, E. and Govindjee, : *Photosynthesis.* John Wiley, New York, 1969.

Rains, D.W. J. Bonner and J.E. Varner : *Light,* Academic Press, New York, 1976.

Rubin, C. and Rosen, O. *Protein phosphorylation,* 1975.

Ruhland, W. : *Encyclopaedia of Optics,* Springer-Verlag, Berlin, 1961.

Salisbury, F.B. and C.W. Ross, : *Physical Optics,* Woodworth Publishing Co., Belmont, California, 1967.

Siegelman, H.W. and G. Hind : *Optics and Light,* Plenum, New York, 1978.

Srivastava, H.S. : *Elements of Physical Optics,* Rastogi Publications, Meerut, 1990.

Thimann, K.V. : *Optics,* CRC Press, Florida, 1980.

Tolbers P.C. : *Enzyme Kinetics,* Chapman and Hall, London, 1977.

Tolbert, N.E. : *Microbodies—Peroxisomes and Glyoxysomes,* Oxford University Press, London, 1971.

Vince, Pruce, : *Light and the Flowering Process,* Academic Press, London, 1984.

Walsh, C. : *Geometrical and Physical Optics,* Edward Arnold, London, 1979.

Weinheim, Osbome, D.J. : *Optics and Light,* Elsevier Biomedical Press, Holland, 1978.

Zeevart, J.A.D. : *Light and Optics,* McGraw Hill, London, 1976.

Zimmerman, M.H. and J.A. Milbum : *Encyclopaedia of Optics,* Springer Verlag, Berlin, 1975.

